MATHEMATICS IN THE TIME
OF THE PHARAOHS

The scribe Ka-Irw-Khufu of the Fourth Dynasty, excavated at Giza. The statue is now owned by the Cairo Museum. Copyright Roger Wood, London.

MATHEMATICS IN THE TIME OF THE PHARAOHS

RICHARD J. GILLINGS

The MIT Press
Cambridge, Massachusetts,
and London, England

121876

This book was designed by The MIT Press Design Department.
It was set in Monotype Baskerville.
Printed on P & S Offset
by Van Rees Book Binding Corp.
and bound in Columbia Milbank Vellum
by Van Rees Book Binding Corp.
in the United States of America.

ISBN 0 262 07045 6 (hardcover)

Library of Congress catalog card number: 74–137469

Dedicated to O. Neugebauer and A. Sachs,
in friendship and gratitude

I wish I knew as much as I thought I knew 10 years ago
O. Neugebauer, 1953

PREFACE xi

1

Introduction 1

2

Hieroglyphic and Hieratic Writing and Numbers 4

3

The Four Arithmetic Operations 11

ADDITION AND SUBTRACTION 11

MULTIPLICATION 16

DIVISION 19

FRACTIONS 20

4

The Two-Thirds Table for Fractions 24

PROBLEMS 61 AND 61B OF THE RHIND MATHEMATICAL PAPYRUS 29

TWO-THIRDS OF AN EVEN FRACTION 32

AN EXTENSION OF RMP 61B AS THE SCRIBE MAY HAVE DONE IT 33

EXAMPLES FROM THE RHIND MATHEMATICAL PAPYRUS OF THE
TWO-THIRDS TABLE 34

5

The G Rule in Egyptian Arithmetic 39

FURTHER EXTENSIONS OF THE G RULE 42

6

The Recto of the Rhind Mathematical Papyrus 45

THE DIVISION OF 2 BY THE ODD NUMBERS 3 TO 101 45

CONCERNING PRIMES 52

FURTHER COMPARISON OF THE SCRIBE'S AND THE COMPUTER'S
DECOMPOSITIONS 53

7

The Recto Continued 71

EVEN NUMBERS IN THE RECTO: 2 ÷ 13 71

MULTIPLES OF DIVISORS IN THE RECTO 74

TWO DIVIDED BY THIRTY-FIVE: THE SCRIBE DISCLOSES HIS METHOD 77

8

Problems in Completion and the Red Auxiliaries 81

USE OF THE RED AUXILIARIES OR REFERENCE NUMBERS 81

AN INTERESTING OSTRACON 86

9

The Egyptian Mathematical Leather Roll   89

THE FIRST GROUP   95

THE SECOND GROUP   95

THE THIRD GROUP   96

THE FOURTH GROUP   98

THE NUMBER SEVEN   99

LINE 10 OF THE FOURTH GROUP   102

THE FIFTH GROUP   102

10

Unit-Fraction Tables   104

UNIT-FRACTION TABLES OF THE RHIND MATHEMATICAL PAPYRUS   106

PROBLEMS 7 TO 20 OF THE RHIND MATHEMATICAL PAPYRUS   109

11

Problems of Equitable Distribution and Accurate Measurement   120

DIVISION OF THE NUMBERS 1 TO 9 BY 10   120

CUTTING UP OF LOAVES   123

SALARY DISTRIBUTION FOR THE PERSONNEL OF THE TEMPLE OF
ILLAHUN   124

12

Pesu Problems   128

EXCHANGE OF LOAVES OF DIFFERENT PESUS   132

13

Areas and Volumes   137

THE AREA OF A RECTANGLE   137

THE AREA OF A TRIANGLE   138

THE AREA OF A CIRCLE   139

THE VOLUME OF A CYLINDRICAL GRANARY   146

THE DETAILS OF KAHUN IV, 3   151

14

Equations of the First and Second Degree   154

THE FIRST GROUP   154

SIMILAR PROBLEMS FROM OTHER PAPYRI   156

THE SECOND AND THIRD GROUPS   158

EQUATIONS OF THE SECOND DEGREE   161

KAHUN LV, 4   162

SUGGESTED RESTORATION OF MISSING LINES OF KAHUN LV, 4, AND
MODERNIZATION OF OTHERS   163

15

Geometric and Arithmetic Progressions   166

GEOMETRIC PROGRESSIONS: PROBLEM 79 OF THE RHIND MATHEMATICAL
PAPYRUS   166

ARITHMETIC PROGRESSIONS: PROBLEM 40 OF THE RHIND
MATHEMATICAL PAPYRUS   170

PROBLEM 64 OF THE RHIND MATHEMATICAL PAPYRUS   173

KAHUN IV, 3   176

16

"Think of a Number" Problems   181

PROBLEM 28 OF THE RHIND MATHEMATICAL PAPYRUS   181

PROBLEM 29 OF THE RHIND MATHEMATICAL PAPYRUS   183

17

Pyramids and Truncated Pyramids   185

THE SEKED OF A PYRAMID   185

THE VOLUME OF A TRUNCATED PYRAMID   187

18

The Area of a Semicylinder and the Area of a Hemisphere   194

19

Fractions of a Hekat   202

20

Egyptian Weights and Measures   207

21

Squares and Square Roots   214

22

The Reisner Papyri: The Superficial Cubit and Scales of
Notation   218

APPENDIX 1

The Nature of Proof   232

APPENDIX 2

The Egyptian Calendar   235

APPENDIX 3

Great Pyramid Mysticism   237

APPENDIX 4

Regarding Morris Kline's Views in *Mathematics, A Cultural Approach*   240

APPENDIX 5

The Pythagorean Theorem in Ancient Egypt   242

APPENDIX 6

The Contents of the Rhind Mathematical Papyrus   243

APPENDIX 7

The Contents of the Moscow Mathematical Papyrus   246

APPENDIX 8

A Papyritic Memo Pad   248

APPENDIX 9

Horus-Eye Fractions in Terms of Hinu: Problems 80, 81 of the Rhind Mathematical Papyrus   249

APPENDIX 10

The Egyptian Equivalent of the Least Common Denominator   251

APPENDIX 11

A Table of Two-Term Equalities for Egyptian Unit Fractions   254

APPENDIX 12

Tables of Hieratic Integers and Fractions, Showing Variations   255

APPENDIX 13

Chronology   260

APPENDIX 14

A Map of Egypt   266

BIBLIOGRAPHY   267

INDEX   277

PREFACE

In 1811, Thomas Cooper, professor of chemistry at Dickinson College, Carlisle, Pennsylvania, and a Britisher by birth, said at his opening lecture, "The history of an art or science is a proper introduction to the study of it; as giving a clear and concise view of the manner in which improvements have been effected; as furnishing due caution against future errors by exhibiting the mistakes of superior minds of olden times, and as rendering merited honor to those who have benefitted mankind by their discoveries."

I have often quoted these words in my own lectures on ancient Egyptian mathematics, but wherever and whenever I have differed from recognized authorities in this field of study, like Eisenlohr, Gillain, Griffith, Sethe, Peet, Struve, Vogel, Chace, Neugebauer, Van der Waerden, and Bruins, I have done so only after due caution and consideration, yet nevertheless with firm and proper conviction. For it was only after carefully examining their translations, their opinions, their solutions and comments, their criticisms, and especially their statements wherein I judged they may have erred, that I have been enabled, I hope, to achieve a truer overall picture of the mathematics of the ancient Egyptians. In other words, I have been privileged to stand on the shoulders of savants and scholars, and the experience has been profoundly rewarding.

R. J. Gillings

*University of New South Wales,
and Turramurra, Australia, 1971*

1 INTRODUCTION

One of the oddest of all the phenomena which come to the attention of students of the history of mathematics is that logarithms were invented by Napier more than a decade before Descartes first conceived the idea of using indices in algebra. This oddity becomes even more striking when we observe that mathematical textbooks today introduce the subject of logarithms by a preliminary study of the *index laws of algebra*, which is pedagogically perhaps the very best way to do it. Chronologically, therefore, the expected order of invention seems to have been reversed; things are the wrong way round!

A second oddity of the history of mathematics was brought to light when the Babylonian clay tablet Plimpton 322 (museum number, Columbia University, New York) was translated by Neugebauer and Sachs in 1945. The translation established beyond any doubt that the Pythagorean theorem was well known to Babylonian mathematicians more than a thousand years before Pythagoras was born. The history books tell us that the Greek mathematician sacrificed an ox to celebrate the discovery of the theorem named after him. Here then is an unrewarded anticipation, for doubtless the name of the famous theorem will remain as a true mumpsimus—"the Pythagorean theorem"—for all time.

Now there is a third oddity in the history of mathematics, which, however, we can clearly understand and explain; it is, indeed, one of the raisons d'être for this book. It is the circumstance that the mathematics, astronomy, and science of the two most ancient of our recorded civilizations, the Egyptian and the Babylonian, have only recently been the subjects of historical research. And the very simple reason for this is that for nearly 3,000 years no one knew what the many extant hieroglyphic and cuneiform writings of these two civilizations meant, nor indeed whether they were writing at all! It was not until Champollion's *Dictionnaire Égyptienne* appeared in 1842 that the Egyptian hieroglyphs were at last deciphered, and not until the latter part of the nineteenth century that the cuneiform writings (beginning with Grotefend in 1802), with their many languages, were deciphered by

Assyriologists, and the secrets so well hidden for centuries—indeed for millennia—at last unfolded.

As a result of these pioneering researches, scholars are now at work in universities, institutes, and museums, transcribing and translating inscriptions from temples and tombs, writings on clay tablets and stelae, hieratic and demotic writings on papyri and ostraca, which had lain unread for years as interesting objects and relics of past civilizations, ancient records whose meaning and significance were only to be guessed or wondered at. Today, scientific and historical journals throughout the world are receiving valuable and informative articles on Egypt and Babylonia to such an extent that they can scarcely cope with the material. In many cases they are months or even years behind their regular publication schedules. The historians' output is setting editors a task that is steadily becoming more formidable.

Of course, some research still proceeds on the histories of the Hindus, the Persians, the Phoenicians, the Hebrews, the Greeks, the Romans, and the Arabic nations, all of whom come much later than the Egyptian culture with which we shall be concerned. The mathematics, astronomy, and science of these other peoples are already well authenticated; deciphered records and scholarly commentary fill many shelves. From the seventeenth century onward, many volumes came from the pens of mathematicians and historians: Moritz Cantor (4 volumes), Johannes Tropfke (7 volumes), Florian Cajori (2 volumes), David Smith (3 volumes), etc. These commentaries and histories took as their starting point the early Greeks, say Thales, about 600 B.C. But knowledge of the earlier civilizations of Egypt and Babylon was not available to these writers, and did not become generally known until the beginning of the present century.

Well may we express our admiration of the wonderful architecture of the Egyptian temples of Karnak and Luxor, at the grandeur and the immensity of the Pyramids and at the construction of their magnificent monuments. Well may we wonder at the government and the economics of a country extending nearly a thousand miles from north to south through which ran the longest river of the then-known world. And well may we marvel at the Egyptians' design of extensive irrigation canals, at their erection of great storage granaries, at the organization of their armies, the building of seagoing ships, the levying and collection of taxes, and at all the thought and effort concomitant with the proper organization of a civilization that existed successfully, virtually unchanged, for centuries longer than that of any other nation in recorded history.

What we today call science and mathematics must have played an important role in the achievement of all this. I am reminded of a piece of wisdom attributed to Arnold Buffum Chace, the principal author of *The Rhind Mathematical Papyrus*:

I venture to suggest that if one were to ask for that single attribute of the human intellect which would most clearly indicate the degree of civilization of a race, the answer would be, the power of close reasoning, and that this power could best be determined in a general way by the mathematical skill which members of the race displayed. Judged by this standard the Egyptians of the nineteenth century before Christ had a high degree of civilization.*

If we accept this thought as one containing a solid measure of truth, then it will surely come as a great surprise to the readers of this history to find that whatever great heights the ancient Egyptians may have achieved scientifically, their mathematics was based on two very elementary concepts. The first was their complete knowledge of the *twice-times table*, and the second, their ability to find *two-thirds of any number*, whether integral or fractional. Upon these two very simple foundations the whole structure of Egyptian mathematics was erected, as we shall see in the following pages.

* A. B. Chace; L. Bull; H. P. Manning; and R. C. Archibald, *The Rhind Mathematical Papyrus*, Vol. 1, Mathematical Association of America, Oberlin, Ohio, 1927, Preface.

2 HIEROGLYPHIC AND HIERATIC WRITING AND NUMBERS

No Egyptian scribe could ever have claimed to be the first man to pick up a mallet and chisel, and to have said to himself, "Now I am going to invent hieroglyphics." He could never have set about carving on a block of stone various figures that would have a special meaning or would convey a message to those who might see it. Neither could it have happened that some intelligent scribe could have been the very first to think of slicing up some Nile River papyrus reeds and, by placing some strips crosswise over others and pressing them flat, invented "paper"; then, bruising the end of a smaller reed, and having dipped it into a pot of "ink," made the dramatic announcement, "Now I am going to write in hieratic characters!"

Neither of these things could have happened like that. The invention of hieroglyphics, which must have come first, took many many years, perhaps centuries. And hieratic writing, the first cursive form of hieroglyphics, developed much later, as a quicker and more convenient way of recording an agreement, conveying a message, or making a calculation with numbers than by the detailed drawing of pictographic hieroglyphs. No one is able to say exactly when writing as we understand it actually began. But with the Egyptians, as with other ancient civilizations, the method used to represent numbers must have been at least easier than writing their phonetically equivalent words. What could be simpler than the scheme shown in Figure 2.1?

Other nations like the Romans, Babylonians, and Chinese thought of something similar for numbers up to 9. The ability to write numbers depended on simple counting, which could have been learned without knowing every separate word for the numbers being written. Most nations adopted a decimal system, no doubt because of the anatomical circumstance that Homo sapiens has ten fingers. Others worked in groups of 5, groups of 20 (the Mayas), or groups of 12 (the Romans); and the Babylonians had a sexagesimal system, which worked partly in groups of 10, and partly in groups of 60. The Egyptian symbols for higher powers of ten are shown in Figure 2.2.

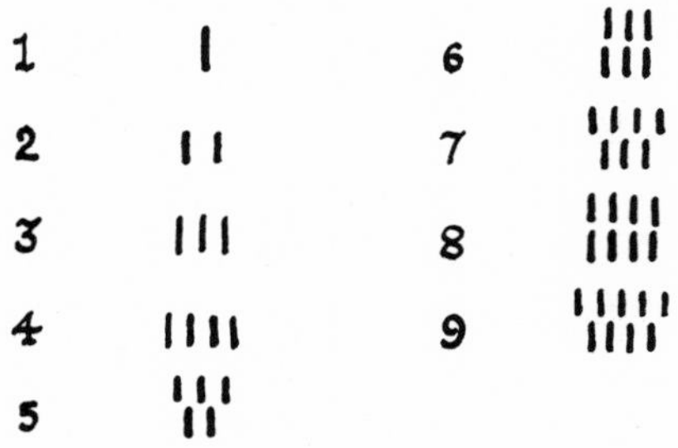

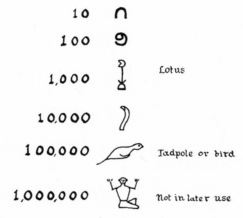

FIGURE 2.1
The earliest Egyptian symbols for the numbers 1 to 9.

FIGURE 2.2
Egyptian symbols for large numbers.

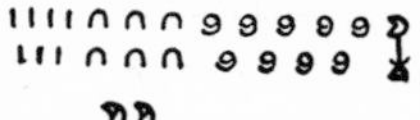

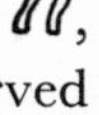

When writing a number in hieroglyphics, sometimes a large number of characters would be required, and often a smaller number would require more characters than a bigger one. Thus the number 1967 in hieroglyphs is ⦙⦙⦙∩∩∩ꝯꝯꝯꝯꝯꝯ𓏦, requiring 23 characters, whereas the number 20,000 is 𝄩𝄩, requiring only 2. It should be remembered that the Egyptians carved or "printed" their hieroglyphs and wrote their cursive hieratic (and later their demotic) forms *from right to left*, just as the Hebrews did, and as the Arabs do today. In contradistinction, however, the Babylonians wrote their cuneiform characters on clay tablets, *from left to right.* Often the hieroglyphic and hieratic writing is written vertically downwards, but still also from right to left. Where the alignment has not been carefully regarded by the scribe, a mathematical papyrus is difficult to follow. In hieroglyphic and hieratic writing, the various birds, reptiles, snakes, and other animals, the scribes, seated or erect, and the human faces, all face the direction from which the writing is coming, when drawn, as most of them are, in profile. It is however an accepted convention that in translation into English or other European language the direction is reversed for convenience, but it is necessary and important, when comparing papyri with the translations, to note whether or not this convention has been observed.

Lack of careful attention to this can be misleading, and even quite wrong, as for example in a recent publication of the LIFE history of mathematics.* There, the two hieroglyphs $\wedge$ and $\wedge$, are stated as meaning *subtract* and *add*, respectively, and the reader is referred to the accompanying illustration where these signs occur. Now the truth is exactly the opposite, and the sense of the mathematical problem,† if examined from the reproduced illustration of the original, is completely changed.

In this present book, if hieroglyphs or hieratic writings are shown or are being discussed as if in situ, then they will be written in the Egyptian fashion from right to left, with hieroglyphs constructed, for example, of animals in profile, facing toward the right. But if we are treating a problem or discussing some mathematical question of Egyptian techniques and methods, we will in translation, for our own convenience, write from left to right.

Now practically everything that comes to our attention will originally have been written in the hieratic script; and, with Egyptian scribes as with present-day handwriters, no two people write the same hand. Some are uniform and some irregular, some are angular and some slope backwards, and as a result the reader needs to acquaint

* David Bergamini and the Editors of LIFE, LIFE *Science Library: Mathematics*, p. 64. © 1965 by TIME Inc. TIME-LIFE International, Netherlands N.V.
† Problem 28 of The Rhind Mathematical Papyrus. See A. B. Chace; L. Bull; H. P. Manning; and R. C. Archibald, *The Rhind Mathematical Papyrus*, Vol. II, Pl. 51.

himself with the idiosyncrasies of different writers. So standard practice among Egyptologists is first of all to transliterate the "cursive" hieratic into "printed" hieroglyphics, and then to translate the hieroglyphics into a modern language. The transliteration is from right to left, and the translation is from left to right. When a restoration is made, in the case of an illegible sign or where there is a lacuna in a poorly preserved papyrus, the custom is to enclose the translation given in square brackets [. . .], and this is done even when the restoration is pretty obvious or even certain. The student of Egyptian and Babylonian mathematics is seldom a competent Egyptologist or Assyriologist as well as being a mathematician, so for translations he is dependent on the skilled specialist in hieratic, demotic, or cuneiform writing. Among such specialists are to be found those who are most competent in some particular branch of the language, as for example, Middle Kingdom hieratic or the Ptolemaic demotic (as on the Rosetta stone), or the cuneiform script (Sumerian, Babylonian, or Akkadian). But there is one powerful factor that is used to advantage by the historian of mathematics. It is that the number systems can all be read and translated by him, even if he cannot understand the words with which the numbers are associated. Indeed he can go a little further. He can with a little industry learn some of the standard symbols commonly associated with mathematical situations—that is, recurrent phrases and specific expressions—that will help in assessing an interpretation provided by the translator, who may not himself be competent in the mathematician's own field. And this is a most important factor with mathematical papyri, because there is one criticism that certainly cannot be leveled at their authors. They were never guilty of circumlocution. The authors of the mathematical papyri were never verbose; indeed, we could often wish that they were sometimes more discursive, that they sometimes went into greater detail. In Figures 2.3–2.5 are given some phrases with their transliterations and translations, that occur in the mathematical papyri. While these examples are by no means exhaustive, they may assist a student who is unacquainted with the hieratic script. They read from right to left, and appear therefore just as they would on a papyrus.

Two or three examples of each are given to exhibit apparent

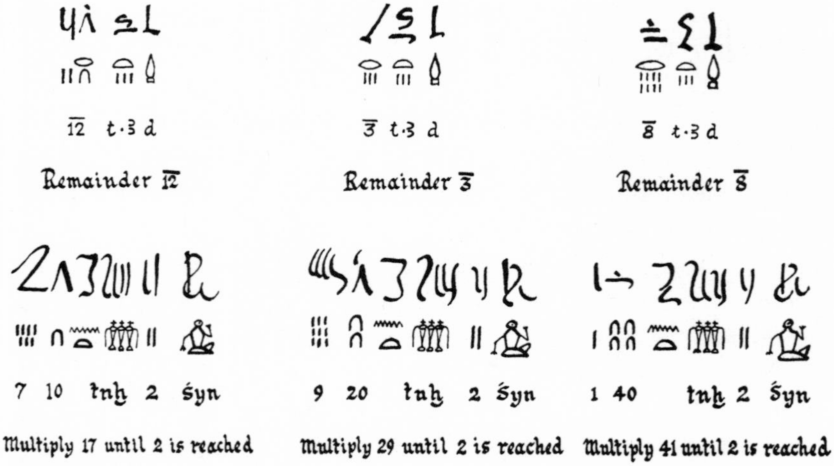

FIGURE 2.3
Examples of Hieratic script, with hieroglyphic transcriptions, phonetic
English transliterations, and English translations. The hieratic characters
and hieroglyphs are written from right to left, and orientable hieroglyphs
(here only "śyn") face toward the right, as in the originals.

differences in cursive handwriting, where ligatures between adjacent
signs may sometimes alter considerably their individual appearance
or character.

FIGURE 2.4
Further examples of hieratic script, transcriptions, transliterations, and
translations. Note that in the examples of vertical writing ("working out")
the orientable figures still face toward the right.

yt yś n pt	yt yś n pt	yt yś n pt
Example of Proof	Example of Proof	Example of Proof
rḥ	rḥ	rḥ
Therefore	Therefore	Therefore
mg	mg	mg
Find	Find	Find
15 dmd	6 dmd	8 dmd
Total 15	Total 6	Total 8
rpḥ y m t·ri	rpḥ y m t·ri	rpḥ y m t·ri
The doing as it occurs	The doing as it occurs	The doing as it occurs
m šš t	m šš t	m šš t
Working out	Working out	Working out

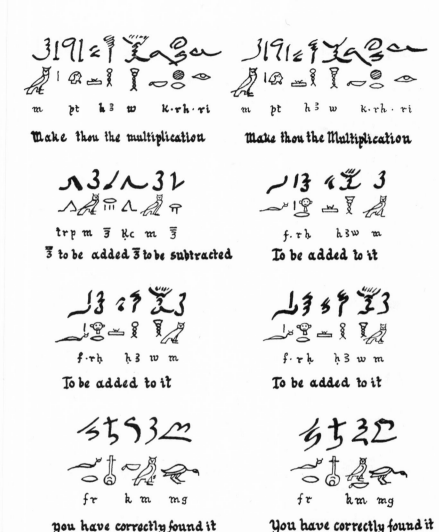

FIGURE 2.5
Further examples of hieratic script, showing the idiosyncratic hand of different scribes.

3 THE FOUR ARITHMETIC OPERATIONS

ADDITION AND SUBTRACTION

Historians of Egyptian mathematics have seldom committed themselves regarding the methods of addition and subtraction employed by the Egyptian scribes; they mostly take for granted the scribes' ability to add or subtract fairly large numbers.* Now we of the twentieth century can count to a million and well past a million, yet *we* have difficulties in addition and subtraction, and we use desk calculators and other aids to computation to overcome them. And we have only to open a modern textbook on the teaching of arithmetic to be surprised at the many different methods for subtraction now being taught. Mankind has always had trouble with these two operations. That is why the abacus was invented centuries ago, and is still being used today in many Asian countries.

In the mathematical papyri that have come down to us, there are many problems proposed and solved which require all four fundamental operations for their solution, and we can quite well deduce the scribes' methods for multiplication and division. But for addition and subtraction methods there are hardly any clues. It would seem that these operations were performed and checked elsewhere by the scribes, and the answers inscribed on the papyri afterwards. A scribal error in this part of their arithmetic is such a rarity that one can be excused for drawing the conclusion that they had tables for additions (and consequently for subtractions), from which they merely read off the answers. If such tables ever existed, however, no copies have come down to us, so that the idea remains purely a conjecture insofar as integers are concerned. For fractions, of course, many varied tables *must* have

---

* For example, see O. Neugebauer, "For ordinary additions and subtractions nothing needs to be said" (*The Exact Sciences in Antiquity*, Harper Torchbooks, Harper, New York, 1962, p. 73). Peet has dismissed the subject by remarking that "people who could count beyond a million had no difficulty about the addition and subtraction of whole numbers" ("Mathematics in Ancient Egypt," *Bulletin of the John Rylands Library*, Vol. 15, No. 2 (Manchester, 1931), p. 412).

			24
			53
			77

			37
			46
			83

			259
			376
			635

FIGURE 3.1

Examples of additions. *Left*, hieratic; *center*, hieroglyphic; *right*, Hindu-Arabic.

existed. The EMLR* is itself very good evidence for the existence of such tables for addition and subtraction of fractions. Today most nations write from *left to right*, and our numbers are so written also; but the values of the digits in our "Hindu-Arabic" decimal system increase in place value from *right to left*. So if we have to perform an addition or a subtraction, we begin with the units column on the right, and work toward the left through the tens, the hundreds, the thousands, and so on. In these calculations, including multiplication, we reverse our direction of writing.

Conversely, as we have seen, the Egyptians wrote both their words and numbers from *right to left*. Of necessity, however, the Egyptian arithmeticians, like ourselves, had to start adding in the opposite direction to that in which they were accustomed to write, so the place value of the Egyptians' digits increases from *left to right*, and the Egyptian system therefore runs widdershins to ours.

Figure 3.1 gives some simple examples of addition comparing Hindu-Arabic, hieroglyphics, and hieratic. In Hindu-Arabic addition, the number combinations $3 + 4 = 7$, $6 + 7 = 13$ and $6 + 9 = 15$, are learned by heart, "look-and-say," so that the addition $(4 + 3)$ is

* Egyptian Mathematical Leather Roll. British Museum, London.

not done by starting with 4, and then counting, 5, 6, 7, and then stopping. This counting method offers itself naturally to one performing additions in hieroglyphics. In hieratic, however, counting does not so lend itself, and we ask ourselves:

(a) *Did they have an additions table?*
(b) *Did they add by simple counting?*
(c) *Did they learn number combinations?*

If tables such as that shown in Figure 3.2 were prepared, they would be equally useful in subtraction, because I am sure the Egyptians did not say to themselves, "From 12 take away 7, answer 5," but rather, "Seven, how many more to make 12? It needs 5," that is, the table would supply the answer.

It is easy to say that multiplication by 10 was simply performed by changing each I to U, each U to 9, each 9 to ?, each ? to ⌐, and

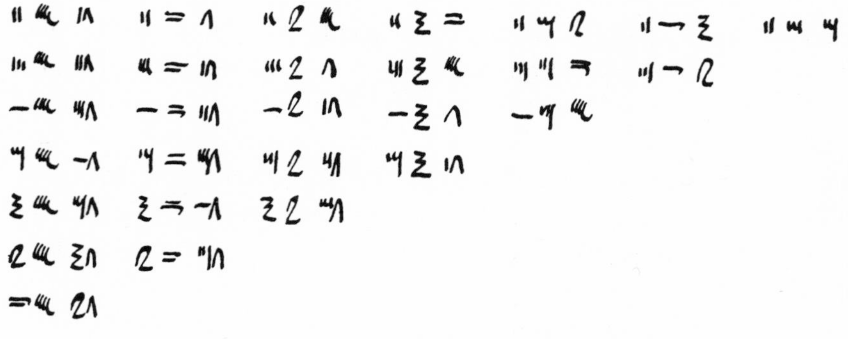

2 9	11	2 8	10	2 7	9	2 6	8	2 5	7	2 4	6	2 3	5
3 9	12	3 8	11	3 7	10	3 6	9	3 5	8	3 4	7		
4 9	13	4 8	12	4 7	11	4 6	10	4 5	9				
5 9	14	5 8	13	5 7	12	5 6	11						
6 9	15	6 8	14	6 7	13								
7 9	16	7 8	15										
8 9	17												

FIGURE 3.2
An addition table in hieratic script that could have been constructed by the scribes, and its Hindu-Arabic translation.

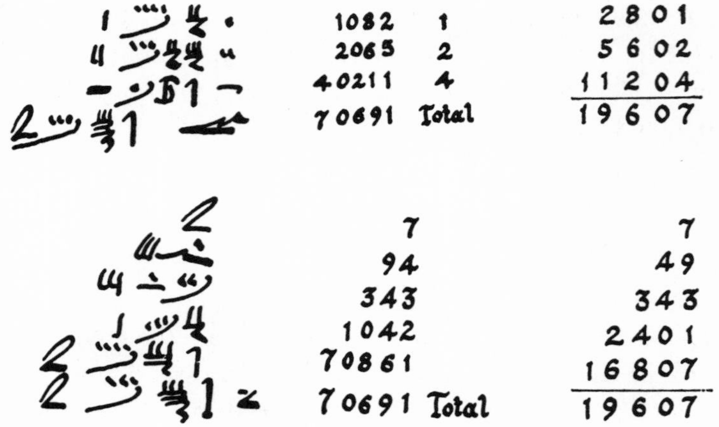

	1082	1		2 8 0 1
	2065	2		5 6 0 2
	4 0211	4		1 1 2 0 4
	7 0 6 91	Total		1 9 6 0 7

	7			7
	94			4 9
	3 4 3			3 4 3
	1 0 4 2			2 4 0 1
	7 0 8 6 1			1 6 8 0 7
	7 0 6 91	Total		1 9 6 0 7

FIGURE 3.3
Two additions from Problem 79 of RMP. *Left,* the hieratic; *center,* the transliteration; *right,* in translation.

so on. But this would be true only of hieroglyphics, and in ordinary everyday business, where these calculations would be needed, hieratic—and later, the quicker demotic—was used by the clerk or scribe, and in such writing this simple transfer of signs does not apply.*

Again, multiplication by continual doubling, performed by a simple duplication of each sign of a number, seems simple enough, even allowing for "carrying," but is feasible only in hieroglyphs, which, as we have seen, were seldom used. We are forced to the conclusion that reference tables were used for the four fundamental operations (which we indicate by $+$, $-$, $\times$, and $\div$); and it is probable that portions of these tables were memorized much as we do today. When we add a column of figures, 9 and 5 immediately suggest 4, 8 and 5 suggest 3, 7 and 5 suggest 2, 6 and 5 suggest 1, and in subtraction these particular pairs suggest the same numbers, so that a kind of elementary theory of numbers begins to arise. There are very many examples of the four operations with unit fractions to be found in the papyri, but very few showing actual addition and subtraction with integers. In Figure 3.3

* See Table 4.6 on hieratic multiplication and division by 10.

are displayed two additions from RMP* Problem 79, the controversial problem thought by some to be the prototype of the Mother Goose rhyme beginning, "As I was going to St. Ives, / I met a man with seven wives. . . ." In the first addition, by chance the units digits are more or less in a vertical line, and (1 + 2 + 4) = 7 simply enough. There are no tens digits. Then for the hundreds there are (800 + 600 + 200), also more or less in line, which must be written as 6 hundreds with one thousand to be carried. What the scribe's thought process for this step was is the point we are doubtful about. Then for the thousands he had (1,000 + 2,000 + 5,000 + 1,000) = 9,000 with nothing to carry, and of course the one ten thousand was merely written down in the total, the whole of which was pushed toward the left, because the symbol for "total," ___, being written first, required more space. His multiplier 7, written by the scribe as (1 + 2 + 4), was not recorded in the last line.

In the second addition, all digits are out of alignment, units, tens, hundreds, etc., but the alternative symbol for "total," ✍, has not pushed the answer out toward the left. In reading this addition greater care is necessary. The scribe had (7 + 9 + 3 + 1 + 7), which, however he arrived at it, is 27, so he put down 7 and carried the 2 tens. Notice the working is from left to right. For the tens he had, (20 + 40 + 40) = 100, so that there are no tens in the answer. The Egyptians had no sign for zero, nor did they even leave a space to indicate "no tens." For the hundreds, he had to add (100 + 300 + 400 + 800), which he finds is 1,000 and 600, (however he did it), so he put down 600 and carried the 1,000. Then he had (1,000 + 2,000 + 6,000) = 9,000 with nothing to carry, which goes into the total with the 10,000.

What we have not enough evidence to decide is whether in adding (100 + 300 + 400 + 800) he counted (on his fingers so to speak), one hundred; two, three, four hundreds; five, six, seven, eight hundreds; then eight and eight hundreds makes sixteen hundreds: or whether he thought of units, adding (1 + 3 + 4 + 8) as 16, and then calling them hundreds. Or did he merely read off the answer from handy prepared tables?

* Rhind Mathematical Papyrus. British Museum.

MULTIPLICATION

It is not uncommon in histories of mathematics to read that Egyptian multiplication was clumsy and awkward, and that this clumsiness and awkwardness was due to the Egyptians' very poor arithmetical notation.

In Ahmes' treatment of multiplication, he seems to have relied on repeated additions.   *Jourdain**

It is remarkable that the Egyptians, who attained so much skill in their arithmetic manipulations, were unable to devise a fresh notation and less cumbersome methods.   *Newman*†

The limitations of this notation made necessary the use of special tables. With such a cumbrous system of fractional notation, calculation was a lengthy process, frequently involving the use of very small fractions. *Sloley*‡

Such a calculus with fractions gave to Egyptian mathematics an elaborate and ponderous character, and effectively impeded the further growth of science.   *Struik*§

If Egyptian multiplication was so clumsy and difficult, how did it come about that these same techniques were still used in Coptic times, in Greek times, and even up to the Byzantine period, a thousand or more years later? No nation, over a period of more than a millennium, was able to improve on the Egyptian notation and methods. The fact is that, despite their notation, the scribes were adepts at solving arithmetic problems and were in fact quite skillful in devising ingenious methods of attack on algebraic and geometric problems as well, so that their successors remained content with what came down to them.

How far have we progressed in multiplication since the times of the ancient Egyptians, or even since Greek and Roman times? What are our grounds for being so critical of Egyptian multiplication, in which

---

* Philip E. B. Jourdain, in *The World of Mathematics*, James R. Newman, ed., Vol. 1, p. 12, Simon and Schuster, New York, 1956.
† *The World of Mathematics*, Vol. 1, p. 172.
‡ R. W. Sloley, in *The Legacy of Egypt*, S. R. K. Glanville, ed., pp. 168f., Oxford University Press, London, 1942 (reprinted 1963).
§ Dirk J. Struik. *A Concise History of Mathematics*, pp. 19f., Dover, New York, 1948.

it was only necessary to use the twice-times tables? In English-speaking countries, at least, as late as the sixteenth century, it was not part of the school curriculum to learn any multiplication tables at all. Samuel Pepys, the famous diarist, educated at St. Paul's School and at Magdalen College, Cambridge University, "an able man of business,"* was secretary to the British Admiralty, and in that position must surely have needed to know how to calculate. But note this entry:

July 4. 1662. . . . Up by 5 o'clock. . . . Comes M. Cooper of whom I intend to learn mathematiques, and do, being with him to-day. After an houres being with him at arithmetique, my first attempt being to learn the multiplication table.†

If a graduate of Cambridge University was just beginning in his thirtieth year to learn the multiplication tables, what are we to suppose the average schoolchild knew of them? In 1542 the Welshman Robert Recorde published *The Grounde of Artes. Teachyng the Worke and Practice of Arithmetike*,‡ in which he shows how to multiply two numbers between 5 and 10.

MULTIPLY 8 BY 7

First set your digits one over the other.

        8
        7
        ———

Then from the uppermost downwards, and from the nethermost upwards, draw straight lines, so that they make a St. Andrew's cross.

        8 ⨯
        7

Then look how many each of them lacketh of 10, and write that against each of them at the end of the line, and that is called the difference.

        8 ⨯ 2
        7 ⨯ 3

I multiply the two differences, saying, "two times three make six," that must I ever set down under the differences.

        8 ⨯ 2
        7 ⨯ 3
            6

---

* Charles J. Finger, *Pepys' Diary*, p. 5.
† *Ibid*, p. 10.
‡ See *The Mathematical Gazette*, London, Vol. XIV, No. 195 (July 1928), pp. 196f.

Take from the other digit, (not from his own), as the
lines of the cross warn me, and that that is left, must I
write under the digits. If I take 2 from 7, or 3 from 8,
(which I will, for all is lyke), and there remaineth 5,
and then there appeareth the multiplication of 8 times
7 to be 56. A chylde can do it.

$$
\begin{array}{cc}
8 & 2 \\
7 & 3 \\
\hline
5 & 6
\end{array}
$$

Compare this technique of the sixteenth century A.D. with that of an
apprentice scribe of the Hyksos period of the Middle Kingdom, more
than 3,000 years earlier.

MULTIPLY 8 BY 7

\1	8
\2	16
\4	32
Totals 7	56

When the Egyptian scribe needed to multiply two numbers, he would
first decide which would be the multiplicand, then he would repeat-
edly multiply this by 2, adding up the intermediate multipliers until
they summed to the original multiplier. For example, to multiply 13
and 7, assume the multiplicand to be 13, doubling thus:

1	13
2	26
4	52

Here he would stop doubling, for he would note that a further step
would give him a multiplier of 8 which is bigger than 7. In this case he
would note that $1 + 2 + 4 = 7$. So he put check marks alongside
these multipliers to indicate this.

\1	13
\2	26
\4	52
Totals 7	91

Adding together those numbers in the right-hand column opposite
check marks, the scribe would thus obtain the final answer. Had he

chosen 7 as the multiplicand, and 13 as the multiplier, his sum would appear as follows:

1	7
2	14
4	28
8	56

Again he would cease doubling at 8, for a further doubling would give 16 which is past 13; then he would note that 1 + 4 + 8 = 13, and so he would put check marks alongside these multipliers.

\1	7
2	14
\4	28
\8	56
Totals 13	91

Since 2 is not checked, he took care not to add in the 14 of the right-hand column, where he has 7 + 28 + 56 = 91. The scribe's mental arithmetic had to be pretty accurate, especially for large multipliers, but of course he could keep a check of his totals on a papyritic scribbling pad. These additions were made easier for the scribe by virtue of a special property of the series

$$1, \quad 2, \quad 4, \quad 8, \quad 16, \quad 32, \quad \ldots \quad ;$$

for any integer can be uniquely expressed as the sum of some of its terms. Thus, for example, 19 = 1 + 2 + 16; 31 = 1 + 2 + 4 + 8 + 16; and 52 = 4 + 16 + 32. We do not know whether or not the scribes were explicitly aware of this but they certainly used it, just as do the designers of a modern electronic computer, and this is surely a somewhat sobering thought.

DIVISION

For the Egyptian scribe, the process of division was closely allied to his method of multiplication. Suppose that he wished to divide 184 by 8. The scribe did not say to himself, "I will divide 8 into 184." He said, "By what must I multiply 8 to get 184?" Thus he had,

1	8
2	16
4	32
8	64
16	128

At this stage he stops multiplying by 2, for his next doubling would give 256, which is well past 184. Now he must do some mental arithmetic or use his memo pad to locate which numbers of the right-hand column will add up to 184. Finding that $8 + 16 + 32 + 128 = 184$, he would place a check mark beside each of these numbers:

	1	8 ✓
	2	16 ✓
	4	32 ✓
	8	64
	16	128 ✓
Totals	23	184

Then he must add the multipliers corresponding to the checked numbers,

$$1 + 2 + 4 + 16 = 23,$$

which is his quotient.

FRACTIONS

When the Egyptian scribe needed to compute with fractions he was confronted with many difficulties arising from the restrictions of his notation. His method of writing numbers did not allow him to write such simple fractions as $3/5$ or $5/9$, because all fractions had to have unity for their numerators (with one exception*). This was because a fraction was denoted by placing the hieroglyph ⬭ ("r," an open mouth) over any integer to indicate its reciprocal. Thus the number 12, written in hieroglyphs as ΙΙ∩, became the fraction $1/12$ when written as ⬭ΙΙ∩. In the hieratic or handwritten form, in which the scribe used a

* The fraction $2/3$. There is some evidence that a special hieroglyph for $3/4$ existed.

reed brush and ink, the open mouth became merely a dot, and $\frac{1}{12}$ would look like $_{\mathbf{II}}\mathbf{\dot{\Lambda}}$. The dot being so much smaller than the "mouth," it was placed over the first digit of the number (here it is 10), and so in reading numbers in hieratic papyri care must be taken not to think that, say, $_{\mathbf{II}}\mathbf{\dot{\Lambda}}$ is $(\frac{1}{10} + 2)$ instead of $\frac{1}{12}$. Such a mistake is even more likely with hundreds or thousands.

Thus all hieroglyphic and hieratic fractions are *unit fractions* (*stammbruchen*), and have unit numerators in translation. The fraction $\frac{3}{4}$ was written by the Egyptians as $\frac{1}{2} + \frac{1}{4}$, $\frac{6}{7}$ was written as, $\frac{1}{2} + \frac{1}{4} + \frac{1}{14} + \frac{1}{28}$, etc., for all fractions of the form p/q. To a modern arithmetician this seems unnecessarily complicated, but we shall see that the Egyptian scribes devised means and rules to meet the difficulties of the method as they arose. The exception to the unit-numerator usage—the fraction $\frac{2}{3}$—was denoted by a special sign: $\mathbf{\Upsilon}$ in hieroglyphics, and $\mathbf{\Upsilon}$ in hieratic. There is no doubt that the Egyptians knew that $\frac{2}{3}$ was the reciprocal of $1\frac{1}{2}$, as the hieroglyph suggests,* for there are many instances, particularly in the RMP, where this relation is specifically shown. Thus:

RMP 33

$$\text{Since } \overline{\overline{3}} \quad \text{of} \quad 42 = 28,$$
$$\text{then } \overline{28} \quad \text{of} \quad 42 = 1\ \overline{2}.$$

RMP 20

$$\overline{\overline{3}} \quad \text{of} \quad \overline{24} = \overline{1\ \overline{2} \times 24},$$
$$= \overline{36}.$$

The numerator 1 of each fraction is here omitted, and $\overline{\overline{3}}$ is $\frac{2}{3}$. In multiplying and dividing fractions, the scribes used the same setting out as they did for integers, but they needed to use various techniques for the different problems that arose. In RMP 2, the fraction $\overline{5}$ is explicitly multiplied by 10 (more often multiplications by 10 were written down at once; see Table 4.3):

---

* In earlier times, the two vertical strokes were often drawn the same length, but this may have been merely lack of care.

l. 1	Do it thus	1		$\overline{5}$		
l. 2		\2	\	$\overline{3}$	$\overline{15}$	
l. 3		4		$\overline{\overline{3}}$	$\overline{10}$	$\overline{30}$
l. 4		\8	\1	$\overline{3}$	$\overline{5}$	$\overline{15}$
l. 5	Totals 10		1	$\overline{\overline{3}}$	$\overline{5}$	$\overline{15}$ $\overline{15}$
l. 6			2			

In line 2 the product $2 \times \overline{5}$ is seen to have been recognized by the scribe as equal to the division $2 \div 5$, for the answer $(\overline{3}\ \overline{15})$ is that given in the RMP Recto* Table, which lists such divisors of 2 (see Chapter 6). Lines 2 and 3 give the double of $\overline{3}$ as $\overline{\overline{3}}$, and $2 \times \overline{15}$ as $(\overline{10}\ \overline{30})$, which, again from the Recto Table, is the result of the division $2 \div 15$. Lines 3 and 4 give the double of $\overline{\overline{3}}$ as $(1\ \overline{3})$, the double of $\overline{10}$ as $\overline{5}$, and the double of $\overline{30}$ as $\overline{15}$. Check marks were put on lines 2 and 4,† and $(\overline{3}\ \overline{15}) + (1\ \overline{3}\ \overline{5}\ \overline{15})$ read off as the answer. The scribe would have recognized this quantity to be equal to 2, by some papyritic jottings on his scribbling pad, or merely by referring to his unit fraction tables of addition.‡

The constant necessity to double fractions in all multiplications gave rise to the construction of the RMP Recto table, illustrated in Table 6.1, where unit fraction equivalents of 2 divided by the odd numbers are recorded for the scribe's easy reference. Many of the simpler equalities were no doubt committed to memory, just as were some of the additions of unit fractions, as shown in the EMLR.

In RMP 9, $\overline{2}\ \overline{14}$ is multiplied by $1\ \overline{2}\ \overline{4}$.

l. 1		\1		\2	$\overline{14}$			
l. 2		\$\overline{2}$		\$\overline{4}$	$\overline{28}$			
l. 3		\$\overline{4}$		\$\overline{8}$	$\overline{56}$			
l. 4	Totals $1\ \overline{2}\ \overline{4}$			$\overline{2}$	$\overline{4}$ $\overline{8}$	$\overline{14}$	$\overline{28}$	$\overline{56}$
l. 5				1				

* The Recto of the RMP is the first portion, dealing with the division of 2 by the odd numbers 3 to 101 (see Chapter 6). The remainder of the RMP is called the Verso.
† Only on one column by the scribe. They are on both columns here for ease of reading.
‡ See Chapter 8.

Line 2 shows the halving of fractions by merely doubling the denominator numbers, and line 3 shows the same. Line 4 shows the totals of both columns, and in line 5, the scribe wrote at once the answer 1. This final addition of six unit fractions was quite possibly done mentally, for $\overline{14}\ \overline{28}\ \overline{56} = \overline{8}$ was a commonplace equality to the scribes,* and the steps

$$
\begin{array}{cccc}
 & \overline{2} & \overline{4} & (\overline{8} \quad \overline{8}) \\
= & \overline{2} & (\overline{4} & \quad \overline{4}) \\
= & (\overline{2} & & \overline{2}) \\
= & & 1 &
\end{array}
$$

would have been quite easy for a competent scribe, even though in writing the details it appears rather long. In more difficult multiplications, as in $7\ \overline{2}\ \overline{4}\ \overline{8} \times 12\ \overline{3}$ (RMP 70), which the scribe showed to be equal to $99\ \overline{2}\ \overline{4}$, he had to refer to his two-thirds table for integers and fractions, to his rule given in RMP 61B, to the Recto Table, and to his 2-term unit fraction tables, and he did it accurately in six lines. It is instructive to calculate $7\frac{7}{8} \times 12\frac{2}{3}$, as we would do it today, and then compare the modern working with that of the ancient scribe. It can be quite an enlightening comparison.

* For example, EMLR 1. 12 is $7\ \overline{14}\ \overline{28} = \overline{4}$, and simple doubling gives $\overline{14}\ \overline{28}\ \overline{56} = \overline{8}$.

4 THE TWO-THIRDS TABLE
FOR FRACTIONS

The one remarkable exception to the fractions with unit numerators was ⅔, which was written with a special sign Υ in hieratic
and $\top$ in hieroglyphic. The scribes used this fraction so freely as an
operator in their multiplications and divisions that one is led to believe
with Peet* that they must have used prepared tables—much of which
they probably knew by heart, as they did their twice-times table. This
two-thirds table was so much a part of a scribe's stock-in-trade that,
were he required to find one-third of a number, he would first find
two-thirds of it and then halve his answer, instead of simply dividing
by 3. This technique was so ingrained that we find it actually being
used for such simple operations as finding one-third of 3 and one-third
of 1! (See RMP 25 and 67.)

In this chapter we examine how the scribes could have produced
such a two-thirds table. Except for the interesting rule given in RMP
61B and a short table giving two-thirds of seventeen simple fractions
in RMP 61, no clear explanations of the Egyptians' methods of obtaining two-thirds of any given number, *integral or fractional*, has come
down to us in any of the mathematical papyri.

A careful check of the problems in the RMP shows that there are 24
examples of the scribe writing $\bar{\bar{3}}$ of integers in one simple operation, of
which the following are illustrations:

$\bar{\bar{3}}$	of	27	is	18,	(RMP Recto)
$\bar{\bar{3}}$	of	365	is	243 $\bar{3}$,	(RMP 66)
$\bar{\bar{3}}$	of	5432	is	3621 $\bar{3}$.	(RMP 33)

There are 14 examples where the scribe wrote down immediately $\bar{\bar{3}}$ of
fractional numbers, such as:

$\bar{\bar{3}}$	of	$\bar{3}$	is	$\bar{6}$ $\overline{18}$,	(RMP 67)
$\bar{\bar{3}}$	of	13 $\overline{23}$	is	8 $\bar{3}$ $\overline{46}$ $\overline{138}$,	(RMP 30)
$\bar{\bar{3}}$	of	7 $\bar{2}$ $\bar{4}$ $\bar{8}$	is	5 $\bar{4}$.	(RMP 70)

* T. E. Peet, "Mathematics in Ancient Egypt," *Bulletin of the John Rylands
Library*, Vol. 15, No. 2 (Manchester, 1931), p. 415.

There are 18 examples where, in order to obtain one-third of a number, the scribe first found $\overline{\overline{3}}$ of it, and then halved his answer. Some examples are

$\overline{\overline{3}}$ of 8:

$$\overline{\overline{3}} \quad \text{of} \quad 8 \quad \text{is} \quad 5\ \overline{3},$$
$$\text{hence,}\ \overline{3}\quad \text{of}\quad 8\quad \text{is}\quad 2\ \overline{3}. \qquad \text{(RMP 43)}$$

$\overline{\overline{3}}$ of 315:

$$\overline{\overline{3}}\quad \text{of}\quad 315\quad \text{is}\quad 210,$$
$$\text{hence,}\ \overline{3}\quad \text{of}\quad 315\quad \text{is}\quad 105. \qquad \text{(RMP 67)}$$

$\overline{\overline{3}}$ of 320:

$$\overline{\overline{3}}\quad \text{of}\quad 320\quad \text{is}\quad 213\ \overline{3},$$
$$\text{hence,}\ \overline{3}\quad \text{of}\quad 320\quad \text{is}\quad 106\ \overline{\overline{3}}. \qquad \text{(RMP 38)}$$

There are a further 17 examples of the scribe writing one-third of a number, omitting the customary intermediate step of first writing its two-thirds, so that we presume he did the necessary halving on an odd piece of papyrus or that he did it mentally by reading directly from a previously prepared two-thirds table. Thus,

$$\overline{3}\quad \text{of}\quad \overline{4}\quad \overline{32}\quad \text{is}\quad \overline{12}\quad \overline{96}, \qquad \text{(RMP 37)}$$

$$\overline{3}\quad \text{of}\quad \overline{4}\quad \overline{53}\quad \overline{106}\quad \overline{212}\quad \text{is}$$
$$\overline{12}\quad \overline{159}\quad \overline{318}\quad \overline{636}. \qquad \text{(RMP 36)}$$

It would be easy to say that

$$\overline{3}\quad \text{of}\quad \overline{4}\quad \overline{32} = \overline{3 \times 4}\quad \overline{3 \times 32}$$
$$= \overline{12}\quad \overline{96},$$

and that the multiplications 3×4 and 3×32 were done in the usual way. But this would be contrary to standard scribal procedure, in which $\overline{\overline{3}}$ was found first and then halved to give $\overline{3}$; thus,

$$\overline{\overline{3}}\quad \text{of}\quad \overline{4}\quad \overline{32} = \overline{6}\quad \overline{48},$$
$$\text{so}\ \overline{3}\quad \text{of}\quad \overline{4}\quad \overline{32} = \overline{12}\quad \overline{96}.$$

And for the second sum, it would also be easy to suppose that $\overline{4}$, $\overline{53}$, $\overline{106}$, and $\overline{212}$ were each multiplied by 3, but the same objections still apply. We should have

$$\overline{\overline{3}} \ \text{ of } \ \overline{4} \quad \overline{53} \quad \overline{106} \quad \overline{212} = \overline{6} \ (\overline{106} \quad \overline{318}) \ \overline{159} \quad \overline{318},$$

$$\text{so } \overline{\overline{3}} \ \text{ of } \ \overline{4} \quad \overline{53} \quad \overline{106} \quad \overline{212} = \overline{12} \ (\overline{212} \quad \overline{636}) \ \overline{318} \quad \overline{636}.$$

(RMP 61B)

But this is not the scribe's answer. The scribe apparently knew that $\overline{212}$ $\overline{636} = \overline{159}$, for he wrote his answer as

$$\overline{12} \quad \overline{159} \quad \overline{318} \quad \overline{636}.$$

He could have gone a step further had he chosen, for

$$\overline{159} \quad \overline{318} = \overline{106},$$
$$\text{and } \overline{318} \quad \overline{636} = \overline{212},$$

so that he could have given either of the simpler answers*

$$\overline{12} \quad \overline{106} \quad \overline{636} \quad \text{or} \quad \overline{12} \quad \overline{159} \quad \overline{212}.$$

However, in RMP 36 these latter refinements are of no moment to the scribe, because these four unit fractions are only a portion of a group of eighteen unit fractions that later have to be added up to give 1. We are here concerned only with how two-thirds and one-third of various numbers could have been found by the scribe, and how he might have prepared his two-thirds table.

Now it would have been quite a simple matter for the scribe to calculate one and one-half times the integers 1 to 10 and set them down like this:

1	$\overline{2}$	times	1 = 1		$\overline{2}$
1	$\overline{2}$		2	3	
1	$\overline{2}$		3	4	$\overline{2}$
1	$\overline{2}$		4	6	
1	$\overline{2}$		5	7	$\overline{2}$
1	$\overline{2}$		6	9	
1	$\overline{2}$		7	10	$\overline{2}$
1	$\overline{2}$		8	12	
1	$\overline{2}$		9	13	$\overline{2}$
1	$\overline{2}$		10	15	

* See Chapters 5, 8.

These numbers form two simple arithmetic progressions, which can be extended by continual addition of 1 $\bar{2}$. Again, this table may be re-written in a slightly different form, in accordance with the prescription

$$\text{if} \quad a \times x = b, \quad \text{then} \quad \bar{b} \times x = \bar{a},$$

which was well known to the scribes, for there are at least 80 examples of it in the RMP Recto alone.* Since the reciprocal of 1 $\bar{2}$ was known to be $\bar{\bar{3}}$, he could have rewritten the table as:

$\bar{\bar{3}}$	of	1	$\bar{2}$	=	1
$\bar{\bar{3}}$		3			2
$\bar{\bar{3}}$		4	$\bar{2}$		3
$\bar{\bar{3}}$		6			4
$\bar{\bar{3}}$		7	$\bar{2}$		5
$\bar{\bar{3}}$		9			6
$\bar{\bar{3}}$		10	$\bar{2}$		7
$\bar{\bar{3}}$		12			8
$\bar{\bar{3}}$		13	$\bar{2}$		9
$\bar{\bar{3}}$		15			10

Now the beginnings of a $\bar{\bar{3}}$ table is emerging, and it may be extended by inserting between each pair of consecutive terms of the progression 1 $\bar{2}$, 3, 4 $\bar{2}$, . . . two equally spaced numbers; for example, between 1 $\bar{2}$ and 3, we insert 2 and 2 $\bar{2}$, and between 3 and 4 $\bar{2}$, we insert 3 $\bar{2}$ and 4, and so on. A similar operation on the progression 1, 2, 3, . . . would give the corresponding series 1, 1 $\bar{3}$, 1 $\bar{\bar{3}}$, 2, 2 $\bar{3}$, 2 $\bar{\bar{3}}$, 3, . . . , so that we can finally list the equalities shown in Table 4.1. This *two-thirds table* would in general suffice for *all* the scribes' needs, and extended up to 100, it would enable them to find ⅔ of any number, integral or fractional, because multiplication and division by 10 was a commonplace operation with them.† Thus in the RMP alone, there are 23 and 20 examples of the immediate multiplication and division,

* Examples are RMP 34: Since $4 \times 1\ \bar{2}\ \bar{4} = 7$, then $\bar{7} \times 1\ \bar{2}\ \bar{4} = \bar{4}$, and RMP 32: Since $1\ \bar{3}\ \bar{4} \times 144 = 228$, then $\overline{228} \times 1\ \bar{3}\ \bar{4} = \overline{144}$.
† See Table 4.6, Multiplication and Division by 10 in the RMP.

TABLE 4.1
The two-thirds table.

$\overline{\overline{3}}$	of	$\overline{2}$	=	$\overline{3}$	(RMP 61)	$\overline{\overline{3}}$	of	8	=	5 $\overline{3}$	(RMP 43)
$\overline{\overline{3}}$		1		$\overline{\overline{3}}$	(RMP 67)	$\overline{\overline{3}}$		8 $\overline{2}$		5 $\overline{\overline{3}}$	
$\overline{\overline{3}}$		1 $\overline{2}$		1		$\overline{\overline{3}}$		9		6	
$\overline{\overline{3}}$		2		1 $\overline{3}$		$\overline{\overline{3}}$		9 $\overline{2}$		6 $\overline{3}$	
$\overline{\overline{3}}$		2 $\overline{2}$		1 $\overline{\overline{3}}$		$\overline{\overline{3}}$		10		6 $\overline{\overline{3}}$	
$\overline{\overline{3}}$		3		2	(RMP 25)	$\overline{\overline{3}}$		10 $\overline{2}$		7	
$\overline{\overline{3}}$		3 $\overline{2}$		2 $\overline{3}$	(RMP 69)	$\overline{\overline{3}}$		11		7 $\overline{3}$	
$\overline{\overline{3}}$		4		2 $\overline{\overline{3}}$		$\overline{\overline{3}}$		11 $\overline{2}$		7 $\overline{\overline{3}}$	
$\overline{\overline{3}}$		4 $\overline{2}$		3		$\overline{\overline{3}}$		12		8	
$\overline{\overline{3}}$		5		3 $\overline{3}$	(RMP 46)	$\overline{\overline{3}}$		12 $\overline{2}$		8 $\overline{3}$	
$\overline{\overline{3}}$		5 $\overline{2}$		3 $\overline{\overline{3}}$		$\overline{\overline{3}}$		13		8 $\overline{\overline{3}}$	
$\overline{\overline{3}}$		6		4	(RMP 39)	$\overline{\overline{3}}$		13 $\overline{2}$		9	
$\overline{\overline{3}}$		6 $\overline{2}$		4 $\overline{3}$		$\overline{\overline{3}}$		14		9 $\overline{3}$	
$\overline{\overline{3}}$		7		4 $\overline{\overline{3}}$		$\overline{\overline{3}}$		14 $\overline{2}$		9 $\overline{\overline{3}}$	
$\overline{\overline{3}}$		7 $\overline{2}$		5	(RMP 70)	$\overline{\overline{3}}$		15		10.	

respectively, by 10, of numbers like 53, 710, 79 $\overline{108}$ $\overline{324}$, 45 $\overline{9}$, 365, and the like.

RMP 40 requires $\overline{\overline{3}}$ of 60, which if not read directly from the extended table, could be found as follows:

$$\overline{\overline{3}} \text{ of } 60 = (\overline{\overline{3}} \text{ of } 6) \times 10$$
$$= 4 \times 10$$
$$= 40.$$

RMP 66 requires the scribe to evaluate $\overline{\overline{3}}$ of 365, the number of days in the Egyptian calendar year:

$$\overline{\overline{3}} \text{ of } 365 = (\overline{\overline{3}} \text{ of } 36 \ \overline{2}) \times 10$$
$$= 24 \ \overline{3} \times 10$$
$$= 240 + 3 \ \overline{3}$$
$$= 243 \ \overline{3}.$$

Although a two-thirds table for unit fractions, such as Table 4.1, has not been preserved for us from the time of the pharaohs, such tables did exist in later times, and some have been preserved. Indeed, quite extensive tables were used as late as the sixth century A.D. The *Akhmîm Papyrus* (J. Baillet, Paris, 1892), in Greek, gives $\overline{\overline{3}}$ of the numbers 2, 3,..., 10; 20, 30,..., 100; 200, 300,..., 1,000; and 2,000, 3,000,..., 10,000.

A similar table, in Coptic and Greek, and ornamented in red, green, and yellow, occurs in Crum's *Catalogue* (London, 1905); this table may date from as late as 1,000 A.D.

PROBLEMS 61 AND 61B OF THE RHIND MATHEMATICAL PAPYRUS

The two problems numbered 61 and 61B by Chace are not really problems but tables of fractions, principally $\bar{\bar{3}}$, $\bar{3}$, and $\bar{2}$, of other fractions. They appear to have been written down as references for the problems that follow them. They are not set down in any real order, although the denominators do get larger as the table proceeds. There is a certain similarity in this regard to the equalities of the EMLR. Some five of the lines appear to be lost, and we can only hazard a guess as to what these lost lines were; but if we rearrange in slightly different order the 17 entries which we do have, we can come close to what these equalities were.

The problem listed as 61B states very clearly the Egyptian rule for finding two-thirds of any odd unit fraction. Chace's translation is:

The making of ⅔ of a fraction uneven. If it is said to thee, What is ⅔ of ⅕? make thou times of it 2, times 6 of it; ⅔ of it this is. Behold does one according to the like for fraction every uneven which may occur.

As we shall see (line 4 of Table 4.2), the scribe knew that this rule also applies to every "even" fraction; but he did not use it very much for even fractions, because he had a much simpler rule for them (line 5). By this latter rule, one merely adds one-half of the number to itself, a technique that is used quite often elsewhere in the RMP. It must have seemed too obvious for the scribe to state this rule in the same detail as for the "uneven" fractions; indeed, as we have observed, the hieroglyph for $\bar{\bar{3}}$ is 𓐍, which suggests that one and one-half times of a fraction was something quite well known.*

There are one or two corrections and erasures in the table of

* There is an interesting line in Problem 19 of the MMP, Moscow Mathematical Papyrus. "Calculate thou with 1 $\bar{2}$ until you find 1. Result $\bar{\bar{3}}$."

TABLE 4.2
Lines taken from Problem 61 of the RMP.

line		of		=		
1	$\overline{\overline{3}}$	of	$\overline{\overline{3}}$	=	$\overline{3}$	$\overline{9}$
2	$\overline{3}$		$\overline{\overline{3}}$		$\overline{6}$	$\overline{18}$ [a]
3	$\overline{\overline{3}}$		$\overline{3}$		$\overline{6}$	$\overline{18}$
4	$\overline{\overline{3}}$		$\overline{6}$		$\overline{12}$	$\overline{36}$
5	$\overline{\overline{3}}$		$\overline{2}$		$\overline{3}$	
6	$\overline{3}$		$\overline{2}$		$\overline{6}$	
7	$\overline{6}$		$\overline{2}$		$\overline{12}$	
8	$\overline{12}$		$\overline{2}$		$\overline{24}$	
9	$\overline{\overline{3}}$		$\overline{9}$		$\overline{18}$	$\overline{54}$ [b]
10	$\overline{4}$		$\overline{5}$		$\overline{20}$	
11	$\overline{\overline{3}}$		$\overline{7}$		$\overline{14}$	$\overline{42}$ [c]
12	$\overline{2}$		$\overline{7}$		$\overline{14}$	
13	$\overline{\overline{3}}$		$\overline{11}$		$\overline{22}$	$\overline{66}$ [d]
14	$\overline{3}$		$\overline{11}$		$\overline{33}$	
15	$\overline{2}$		$\overline{11}$		$\overline{22}$	
16	$\overline{4}$		$\overline{11}$		$\overline{44}$	
17	$\overline{\overline{3}}$		$\overline{5}$		$\overline{10}$	$\overline{30}$.

[a] The scribe could have checked from the Recto, 2 divided by the odd numbers, for $\overline{3}$ of $\overline{3}$ is the same as $2 \div 9$, which is there proved to be $\overline{6}$ $\overline{18}$. See Chapter 6.

[b] Again in the Recto, $2 \div 27$ is given as $\overline{18}$ $\overline{54}$. This line is repeated in the papyrus.

[c] As previously, the Recto gives $2 \div 21 = \overline{14}$ $\overline{42}$.

[d] From the Recto, $2 \div 33 = \overline{22}$ $\overline{66}$.

Problem 61, and line 9 is duplicated; the sequence as it now stands (Table 4.2) contains no scribal errors.

From the equalities given in Table 4.2, we now make a table of two-thirds of *every* unit fraction from $\overline{2}$ to $\overline{12}$, in which we have to include four lines not given by the scribe (see Table 4.3).

The table for one-third of every fraction from $\overline{2}$ to $\overline{12}$ (Table 4.4) contains only three of the scribe's entries. Indeed, the scribe did not need an extensive one-third table, because each entry could be found by merely doubling the corresponding entries in the two-thirds table. Five equalities of the one-half table (Table 4.5) occur in RMP 61. The scribe was also aware that every equality in these tables may be stated in reverse; for example, $\overline{\overline{3}}$ of $\overline{3}$ and $\overline{3}$ of $\overline{\overline{3}}$ are the same ($\overline{6}$ $\overline{18}$). He also was aware that any one equality would produce another by

TABLE 4.3
Two-thirds of unit fractions.

$\overline{3}$	of	$\overline{2}$	=	$\overline{4}$	$\overline{12}$	=	$\overline{3}$
$\overline{3}$		$\overline{3}$		$\overline{6}$	$\overline{18}$		
$\overline{3}$		$\overline{4}$		$\overline{8}$	$\overline{24}$		$\overline{6}$
$\overline{3}$		$\overline{5}$		$\overline{10}$	$\overline{30}$		
$\overline{3}$		$\overline{6}$		$\overline{12}$	$\overline{36}$		$\overline{9}$
$\overline{3}$		$\overline{7}$		$\overline{14}$	$\overline{42}$		
$\overline{3}$		$\overline{8}$		$\overline{16}$	$\overline{48}$		$\overline{12}$
$\overline{3}$		$\overline{9}$		$\overline{18}$	$\overline{54}$		
$\overline{3}$		$\overline{10}$		$\overline{20}$	$\overline{60}$		$\overline{15}$
$\overline{3}$		$\overline{11}$		$\overline{22}$	$\overline{66}$		
$\overline{3}$		$\overline{12}$		$\overline{24}$	$\overline{72}$		$\overline{18}$.

TABLE 4.4
One-third of unit fractions.

$\overline{3}$	of	$\overline{2}$	=	$\overline{8}$	$\overline{24}$	=	$\overline{6}$
$\overline{3}$		$\overline{3}$		$\overline{12}$	$\overline{36}$		$\overline{9}$
$\overline{3}$		$\overline{4}$		$\overline{16}$	$\overline{48}$		$\overline{12}$
$\overline{3}$		$\overline{5}$		$\overline{20}$	$\overline{60}$		$\overline{15}$
$\overline{3}$		$\overline{6}$		$\overline{24}$	$\overline{72}$		$\overline{18}$
$\overline{3}$		$\overline{7}$		$\overline{28}$	$\overline{84}$		$\overline{21}$
$\overline{3}$		$\overline{8}$		$\overline{32}$	$\overline{96}$		$\overline{24}$
$\overline{3}$		$\overline{9}$		$\overline{36}$	$\overline{108}$		$\overline{27}$
$\overline{3}$		$\overline{10}$		$\overline{40}$	$\overline{120}$		$\overline{30}$
$\overline{3}$		$\overline{11}$		$\overline{44}$	$\overline{132}$		$\overline{33}$
$\overline{3}$		$\overline{12}$		$\overline{48}$	$\overline{144}$		$\overline{36}$.

TABLE 4.5
One-half of unit fractions.

$\overline{2}$	of	$\overline{2}$	=	$\overline{4}$
$\overline{2}$		$\overline{3}$		$\overline{6}$
$\overline{2}$		$\overline{4}$		$\overline{8}$
$\overline{2}$		$\overline{5}$		$\overline{10}$
$\overline{2}$		$\overline{6}$		$\overline{12}$
$\overline{2}$		$\overline{7}$		$\overline{14}$
$\overline{2}$		$\overline{8}$		$\overline{16}$
$\overline{2}$		$\overline{9}$		$\overline{18}$
$\overline{2}$		$\overline{10}$		$\overline{20}$
$\overline{2}$		$\overline{11}$		$\overline{22}$
$\overline{2}$		$\overline{12}$		$\overline{24}$.

multiplying the appropriate number on each side by 2, 3, 4, ... Lines 6, 7, and 8 of Table 4.2 show this very clearly:

$$1.6 \quad \overline{3} \quad \text{of} \quad \overline{2} = \overline{6},$$
$$1.7 \quad \overline{6} \quad \text{of} \quad \overline{2} = \overline{12}.$$
$$1.8 \quad \overline{12} \quad \text{of} \quad \overline{2} = \overline{24}.$$

Each line is obtained from the preceding on multiplying the first and last numbers by 2. Again in lines 3 and 4, the same is observed:

$$1.3 \quad \overline{\overline{3}} \quad \text{of} \quad \overline{3} = \overline{6} \quad \overline{18}$$
$$1.4 \quad \overline{\overline{3}} \quad \text{of} \quad \overline{6} = \overline{12} \quad \overline{36}$$

Also,

$$1.15 \quad \overline{2} \quad \text{of} \quad \overline{11} = \overline{22}.$$
$$1.16 \quad \overline{4} \quad \text{of} \quad \overline{11} = \overline{44}.$$

That each of the $\overline{\overline{3}}$, $\overline{3}$, and $\overline{2}$ tables could be extended as far as he desired was clear to the scribe, and in fact he was aware that further tables for $\overline{4}, \overline{5}, \overline{6}, \ldots$ could as easily be drawn up, because many of the entries would already have been included in the tables that he already had.

TWO-THIRDS OF AN EVEN FRACTION

The following examples are from the RMP, where the scribe used the rule for finding $\overline{\overline{3}}$ of an even unit fraction, by adding to the number its half:

$\overline{\overline{3}}$ of	$\overline{2} =$	$\overline{3}$	Problems 16, 61, 69, 70.
$\overline{\overline{3}}$	$\overline{4}$	$\overline{6}$	8, 32, 70.
$\overline{\overline{3}}$	$\overline{6}$	$\overline{9}$	18, 32, 42, 61.
$\overline{\overline{3}}$	$\overline{8}$	$\overline{12}$	70.
$\overline{\overline{3}}$	$\overline{12}$	$\overline{18}$	19, 32.
$\overline{\overline{3}}$	$\overline{18}$	$\overline{27}$	42.
$\overline{\overline{3}}$	$\overline{24}$	$\overline{36}$	20.
$\overline{\overline{3}}$	$\overline{56}$	$\overline{84}$	33.
$\overline{\overline{3}}$	$\overline{114}$	$\overline{171}$	32.
$\overline{\overline{3}}$	$\overline{228}$	$\overline{342}$	32.
$\overline{\overline{3}}$	$\overline{776}$	$\overline{1164}$	33.

AN EXTENSION OF RMP 61B AS THE SCRIBE MAY HAVE DONE IT

Since *two-thirds of any odd (or even) fraction is equal to 2 times it plus 6 times it*, why not restate the rule as follows? The scribes may well have done this. *One-third of any odd (or even) fraction is equal to 4 times it plus 12 times it*. Then what do we get?

$$\overline{3} \quad \text{of} \quad \overline{1} = \overline{4} \quad \overline{12}$$
$$\overline{3} \qquad\qquad \overline{2} \quad\; \overline{8} \quad\; \overline{24}$$
$$\overline{3} \qquad\qquad \overline{3} \quad \overline{12} \quad \overline{36}$$
$$\overline{3} \qquad\qquad \overline{4} \quad \overline{16} \quad \overline{48}$$
$$\overline{3} \qquad\qquad \overline{5} \quad \overline{20} \quad \overline{60}$$
$$\overline{3} \qquad\qquad \overline{6} \quad \overline{24} \quad \overline{72}$$

. . . .

But from simple first principles, we also have

$$\overline{3} \quad \text{of} \quad \overline{1} = \overline{3}$$
$$\overline{3} \qquad\qquad \overline{2} \quad\; \overline{6}$$
$$\overline{3} \qquad\qquad \overline{3} \quad\; \overline{9}$$
$$\overline{3} \qquad\qquad \overline{4} \quad \overline{12}$$
$$\overline{3} \qquad\qquad \overline{5} \quad \overline{15}$$
$$\overline{3} \qquad\qquad \overline{6} \quad \overline{18}$$

. . . .

Then we have the 2-term unit-fraction equalities:

$$\overline{4} \quad\; \overline{12} = \overline{3}$$
$$\overline{8} \quad\; \overline{24} = \overline{6}$$
$$\overline{12} \quad \overline{36} = \overline{9}$$
$$\overline{16} \quad \overline{48} = \overline{12}$$
$$\overline{20} \quad \overline{60} = \overline{15}$$
$$\overline{24} \quad \overline{72} = \overline{18}$$

. . . .

This is the table of equalities developed from the generator (1, 3) that we will meet again when we discuss the G rule in Chapter 5.

By applying the scribe's rule of RMP 61B to the smaller odd numbers, we have

$$
\begin{array}{llllll}
\overline{3} & \text{of} & \overline{3} & = & \overline{6} & \overline{18} \\
\overline{3} & & \overline{5} & & \overline{10} & \overline{30} \\
\overline{3} & & \overline{7} & & \overline{14} & \overline{42} \\
\overline{3} & & \overline{9} & & \overline{18} & \overline{54} \\
\overline{3} & & \overline{11} & & \overline{22} & \overline{66}
\end{array}
$$

. . . .

Now again from first principles, $\overline{3}$ of $\overline{3}$ can be expressed as $2 \div 9$, $\overline{3}$ of $\overline{5}$ as $2 \div 15$, and so on, so that we can rewrite the preceding table as

$$
\begin{array}{llll}
2 \div 9 & = & \overline{6} & \overline{18} \\
2 \div 15 & = & \overline{10} & \overline{30} \\
2 \div 21 & = & \overline{14} & \overline{42} \\
2 \div 27 & = & \overline{18} & \overline{54} \\
2 \div 33 & = & \overline{22} & \overline{66}
\end{array}
$$

. . . .

These are *exactly* the entries to be found in the Recto of the RMP, and there are just 16 of them, those where the divisors are multiples of 3. It is thus possible that the scribe checked his answers to these particular divisions by means of this extension of RMP 61B.

EXAMPLES FROM THE RHIND MATHEMATICAL PAPYRUS OF THE TWO-THIRDS TABLE

Considering the complexities of finding $\frac{2}{3}$ of some numbers, which the scribe apparently did in his head, it is surprising to see the working-out of one-third of 3 and of 1, found in all detail in Problems 25, 67.

PROBLEM 25

$\overline{3}$ of $3 = 1$.

$$
\begin{array}{ll}
1 & 3 \\
\overline{3} & 2 \\
\overline{3} & 1.
\end{array}
$$

PROBLEM 67

$\overline{3}$ of $1 = \overline{3}$.

$$
\begin{array}{ll}
1 & 1 \\
\overline{3} & \overline{3} \\
\overline{3} & \overline{3}
\end{array}
$$

PROBLEM 32

$\overline{\overline{3}}$ of 1 $\overline{3}$ $\overline{4}$ = $\overline{\overline{2}}$ $\overline{36}$.

1		1	$\overline{3}$	$\overline{4}$	
$\overline{\overline{3}}$		$\overline{\overline{3}}$ ($\overline{6}$ $\overline{18}$)	$\overline{6}$		
$\overline{\overline{3}}$		$\overline{\overline{3}}$ ($\overline{6}$ $\overline{6}$)	$\overline{18}$		(rearranging)
$\overline{\overline{3}}$		$\overline{\overline{3}}$	$\overline{3}$	$\overline{18}$	
$\overline{\overline{3}}$		1	$\overline{18}$		
$\overline{\overline{3}}$		$\overline{\overline{2}}$	$\overline{36}$.		

In Problem 32 the scribe does not show all the intermediate steps given above; his work appears simply as follows:

1		1 $\overline{3}$ $\overline{4}$
$\overline{\overline{3}}$		1 $\overline{18}$
$\overline{\overline{3}}$		$\overline{\overline{2}}$ $\overline{36}$.

Without any explanatory steps, the scribe writes at once two-thirds of numbers, both integral and fractional, on at least 38 occasions in the RMP. Some of these appear quite difficult, as the following examples will show.

PROBLEM 42

1		8 $\overline{\overline{3}}$ $\overline{6}$ $\overline{18}$
$\overline{\overline{3}}$		5 $\overline{\overline{3}}$ $\overline{6}$ $\overline{18}$ $\overline{27}$.

PROBLEM 33

1		16 $\overline{56}$ $\overline{679}$ $\overline{776}$
$\overline{\overline{3}}$		10 $\overline{\overline{3}}$ $\quad$ $\overline{84}$ $\overline{1358}$ $\overline{4074}$ $\overline{1164}$.

PROBLEM 32

1		1 $\overline{6}$ $\overline{12}$ $\overline{114}$ $\overline{228}$
$\overline{\overline{3}}$		$\overline{\overline{3}}$ $\overline{9}$ $\overline{18}$ $\overline{171}$ $\overline{342}$.

The comparative ease and facility with which two-thirds of a complicated fraction was achieved is astounding to us—more so, if we should attempt to check the scribes' accuracy by our modern methods. Scribal errors were rare!

RMP 61B gives the Egyptian rule for finding $\overline{\overline{3}}$ of any odd fraction, which is to take twice the fraction plus six times the fraction. The

TABLE 4.6
Examples from the RMP of multiplication and division by 10. As can be
seen from the hieratic, all answers were given with no working.

$2 \div 15$	$15 \div 10$	$=$	$1\ \bar{2}$		
$2 \div 23$	23×10	$=$	230		
$2 \div 47$	47×10	$=$	470		
$2 \div 53$	53×10	$=$	530		
$2 \div 61$	61×10	$=$	610		
$2 \div 71$	71×10	$=$	710		
$2 \div 79$	79×10	$=$	790		
$2 \div 89$	89×10	$=$	890		
Problem 21	$15 \div 10$	$=$	$1\ \bar{2}$		
" 22	$30 \div 10$	$=$	3		
" 29	$10 \div 10$	$=$	1		
" 30	$13\ \overline{23} \div 10$	$=$	$1\ \bar{5}\ \overline{10}\ \overline{230}$		
" 35	$3\ \bar{3} \div 10$	$=$	$\bar{3}$		
" 35	$320 \div 10$	$=$	32		
" 39	4×10	$=$	40		
" 41	64×10	$=$	640		
" 41	$960 \div 10$	$=$	96		
" 42	$79\ \overline{108}\ \overline{324} \times 10$	$=$	$790\ \overline{18}\ \overline{27}\ \overline{54}\ \overline{81}$		
" 43	$10\ \bar{\bar{3}} \times 10$	$=$	$106\ \bar{\bar{3}}$		
" 44	10×10	$=$	100		

TABLE 4.6 (*continued*)

Problem 43	$455\,\bar{9} \div 10$	=	$45\,\bar{2}\,\overline{90}$		
"	44	100×10	=	1000	
"	44	$1500 \div 10$	=	150	
"	44	75×10	=	750	
"	44	$150 \div 10$	=	15	
"	46	25×10	=	250	
"	49	1000×10	=	10000	
"	49	10000×10	=	100000	
"	49	$10000 \div 10$	=	1000	
"	55	$5 \div 10$	=	$\bar{2}$	
"	66	$365 \div 10$	=	$36\,\bar{2}$	
"	69	$3\,\bar{2} \times 10$	=	35	
"	70	100×10	=	1000	
"	70	$\overline{16}\,\overline{64} \times 10$	=	$\bar{2}\,\bar{4}\,\overline{32}$	
"	70	$\bar{5} \times 10$	=	2	
"	70	$\bar{2}\,\bar{4}\,\overline{32} \times 10$	=	$7\,\bar{2}\,\bar{4}\,\overline{16}$	
"	72	$35 \div 10$	=	$3\,\bar{2}$	
"	73	15×10	=	150	
"	76	$2\,\bar{2} \times 10$	=	25	
"	82	$1\,\bar{4} \times 10$	=	$12\,\bar{2}$	

scribes could have found out this rule from simpler 2-term unit fractions, thus,

$$
\begin{aligned}
\text{since } \bar{3}\ \bar{3} \quad &= (\bar{6}\ \bar{6})\ (\bar{6}\ \bar{6}) \\
&= (\bar{6}\ \bar{6}\ \quad \bar{6})\,\bar{6} \\
&= \quad\ \bar{2} \qquad \bar{6}
\end{aligned}
$$

$$
\text{then} \quad \bar{\bar{3}} \quad = \quad \bar{2} \qquad \bar{6},
$$

$$
\text{or} \quad \bar{\bar{3}} \text{ of } \bar{n} = \quad \overline{2n} \qquad \overline{6n}.
$$

As we have seen, this $\bar{\bar{3}}$ rule applies to all n, whether odd or even; but the scribes see no good reason for applying it to even numbers, because $\bar{\bar{3}}$ of $\overline{2n}$ is obviously equal to $\bar{\bar{3}}$ of $\bar{n}$, and this by simple multiplication is $\overline{3n}$.

In attempting to establish that the Egyptian arithmeticians must have been aware of what I have called the G rule, I will have to exhibit many examples of 2-term equalities. By far the greatest number of instances where the scribes appear to have used the G rule in one form or another occur in the Recto and the 87 problems of the RMP, in the MMP, and the KP.* However, it will be convenient to choose from the equalities of the EMLR for purposes of illustration and demonstration, even though the EMLR itself will not be considered until Chapter 9. The easy references and the small numerical magnitudes lend themselves to such convenient analysis that I cannot ignore the opportunity provided by the EMLR.

The G rule is, to my knowledge, nowhere *explicitly* stated in the extant Egyptian papyri. However, I hope to show, from the mathematical evidence at our disposal, that the scribes probably knew and often used this rule. First, look at these 10 lines from the EMLR:

l. 11	$\overline{9}$	$\overline{18} = \overline{6}$
l. 13	$\overline{12}$	$\overline{24} = \overline{8}$
l. 19	$\overline{24}$	$\overline{48} = \overline{16}$
l. 20	$\overline{18}$	$\overline{36} = \overline{12}$
l. 21	$\overline{21}$	$\overline{42} = \overline{14}$
l. 22	$\overline{45}$	$\overline{90} = \overline{30}$
l. 23	$\overline{30}$	$\overline{60} = \overline{20}$
l. 24	$\overline{15}$	$\overline{30} = \overline{10}$
l. 25	$\overline{48}$	$\overline{96} = \overline{32}$
l. 26	$\overline{96}$	$\overline{192} = \overline{64}$.

An intelligent scribe would certainly notice a certain simple relation existing between the three terms of each of these equalities. The expression of this relation is the G rule. In modern mathematical terms we may state it as follows:

G rule: If one unit fraction is double another then their sum is a

* Kahun Papyrus. British Museum, London.

different unit fraction *if and only if* the larger denominator is divisible by 3. The quotient of the division is the unit fraction of the sum.

But if such a rule were ever expressed by an Egyptian scribe, it would have been much terser, probably something like this:

For adding 2 fractions, if one number is twice the other, divide it by 3.

Line 11 of the EMLR illustrates the G rule:

$$\overline{9} \quad \overline{18} = \overline{6}.$$

This could be written down immediately, because 18 is twice 9, and then $18 \div 3$ gives 6, which is the unit fraction sum. The other nine lines also obey the rule in the same way.

The G rule would have been of further usefulness to the scribes because, if the larger fraction does not give an integer when divided by 3, then a single-term unit-fraction sum is not possible. Take for example the addition of the two unit fractions, $\overline{5}$ and $\overline{10}$:

$$\overline{5} \quad \overline{10} = ?$$

Although 10 is twice 5, $10 \div 3 = 3\ \overline{3}$ which is not an integer, so that however $\overline{5} + \overline{10}$ may be otherwise expressed, it certainly cannot be written as a single unit fraction.

The G rule could have been extended by the scribes who observed, for instance, the equality in line 3 of the EMLR. This is

$$\overline{4} \quad \overline{12} = \overline{3}.$$

One notices in this case that one of the paired unit fractions is 3 times the other ($\overline{4}$ is 3 times $\overline{12}$, or, as the Egyptians would have phrased it, 12 is 3 times 4); adding 1 to this multiplier (by analogy with the G rule) and dividing into $\overline{12}$, we obtain the sum $\overline{3}$. Indeed the scribes were able to go further. Let us look at lines 1 and 2, the first two entries in the EMLR. They are

$$\overline{10} \quad \overline{40} = \overline{8},$$
$$\overline{5} \quad \overline{20} = \overline{4}.$$

It is a fair assumption that the scribe responsible for the EMLR put them together because they are similar equalities. In each case, the larger number is 4 times the other, and by adding 1 as before, one obtains 5, which divided in turn into 40 and 20, respectively, gives the sums $\bar{8}$ and $\bar{4}$. Thus the scribes could have arrived by induction at a generalization that we can state in modern terms as:

Extension of the G Rule: If one of two unit fractions is K times the other, then their sum is found by dividing $(K + 1)$ into the larger number, providing the answer is an integer.

Further, if the answer is not an integer, then a unit fraction sum is impossible, and thus the rule becomes doubly useful.

I am sure that, over the many centuries in which the scribes added and subtracted unit fractions, they often observed and made use of this rule, perhaps not exactly in the form that we have expressed it. But the principle must have been known, even though no explicit statement of it is given in any of the extant Egyptian papyri. One thing is certain. Anyone working today in Egyptian mathematics will find the rule immeasurably useful in checking computations. All of the 2-term equalities shown in Table 5.1 appear to have been written down mentally by the scribe of the RMP; of course, they may have been checked from prepared tables.

TABLE 5.1
Instances of the scribe's probable use of the extended G rule in the RMP.

Location	Equality			
Recto, 2 ÷ 17, 37, 43, 59, 73, 79, 83, 89, 95, 101. Verso, Problems 4, 5, 16, 38, 56, 66, 69.	$\bar{3}$	$\bar{6}$	=	$\bar{2}$.
Recto, 2 ÷ 19, 23, 95.	$\bar{6}$	$\overline{12}$	=	$\bar{4}$.
Recto, 2 ÷ 59. Verso, Problems 17, 18, 67.	$\bar{9}$	$\overline{18}$	=	$\bar{6}$.
Verso, Problems 30, 35.	$\overline{15}$	$\overline{30}$	=	$\overline{10}$.
Recto, 2 ÷ 59, 95. Verso, Problem 33.	$\bar{4}$	$\overline{12}$	=	$\bar{3}$.
Recto, 2 ÷ 29, 37, 41.	$\bar{8}$	$\overline{24}$	=	$\bar{6}$.
Recto, 2 ÷ 31, 67, 73, 83, 89.	$\bar{5}$	$\overline{20}$	=	$\bar{4}$.
Recto, 2 ÷ 61, 71.	$\overline{10}$	$\overline{40}$	=	$\bar{8}$.
Recto, 2 ÷ 43.	$\bar{7}$	$\overline{52}$	=	$\bar{6}$.
Verso, Problem 33.	$\overline{14}$	$\overline{84}$	=	$\overline{12}$.
Recto, 2 ÷ 47, 53, 79. Verso, Problems 4, 5, 56.	$\overline{10}$	$\overline{15}$	=	$\bar{6}$.

I will now establish the standard equality, $\bar{7}\ \overline{14}\ \overline{28} = \bar{4}$ as an exercise in the application of the G rule. First multiply the left-hand side by, 2, 3, . . . :

1. 1		$\bar{7}$	$\overline{14}$	$\overline{28}$
1. 2	multiply by 2	$\overline{14}$	$\overline{28}$	$\overline{56}$
1. 3	,, 3	$\overline{21}$	$\overline{42}$	$\overline{84}$
1. 4	,, 4	$\overline{28}$	$\overline{56}$	$\overline{112}$
1. 5	,, 5	$\overline{35}$	$\overline{70}$	$\overline{140}.$

. . . .

We choose the third line because it is the only one with numbers divisible by 3. Then we have

$$
\begin{aligned}
\overline{21}\ \overline{42}\ \overline{84} &= (\overline{21}\ \overline{42})\ \overline{84} \\
&= (\overline{14}\qquad \overline{84}) \qquad [42 \div 3 = 14.] \\
&= \qquad \overline{12}. \qquad\qquad [84 \div 7 = 12.]
\end{aligned}
$$

Then, dividing both sides of $\overline{21}\ \overline{42}\ \overline{84} = \overline{12}$ by 3 gives

$$\bar{7}\ \overline{14}\ \overline{28} = \bar{4}.$$

The repeated multiplication illustrated here is an often-fruitful technique for casting many-term fractions into a form amenable to computation with the G rule. Here, although 7 divides 14, giving 2, yet 3 does not divide 14, and again, although 14 divides 28, giving 2, yet 3 does not divide 28; but multiplication by 3 gives the sum $\overline{21}\ \overline{42}\ \overline{84}$, upon which the G rule may operate.

FURTHER EXTENSIONS OF THE G RULE

There are many examples in the arithmetical calculations of the scribes where the subtractive concept with unit fractions predominates over the additive concept. It was necessary then for them only to look at the tables they already had for the addition of unit fractions (in, e.g., the EMLR), from a slightly different point of view, and perhaps even rewrite them, to produce a table for the difference of unit fractions. One must remember of course that they had no specific signs for plus and minus; mere juxtaposition of numbers meant "add," and if subtraction was meant, then the words in hieroglyphs or hieratic had to be written. Although no such tables for the *difference* of unit fractions have come down to us, it is instructive to examine how they may have been derived from the *addition* tables, and to see what they would have looked like. We introduce plus and minus signs in these equalities for clarity, and, anticipating Chapter 10, we take first the table whose generator is (1, 2)—that is, the table formed of consecutive integral multiples of 1 and 2—and omit for simplicity the bars over the numbers, for they are all unit fractions.

GENERATOR (1, 2)

$$3 + 6 = 2$$
$$6 + 12 = 4$$
$$9 + 18 = 6$$
$$\cdots$$

We see that each entry in this table is a G-rule equality. And now we rewrite the table as follows:

GENERATOR (1, 3)

$$2 - 6 = 3$$
$$4 - 12 = 6$$
$$6 - 18 = 9$$
$$\cdots$$

Now we can see that, with one simple alteration, the G rule applies to the *subtraction* of unit fractions. Thus, for the first equality, one says 2 into 6 is 3, (3 − 1) is 2, and 2 into 6 is 3, so that the difference of the unit fractions $\frac{1}{2}$ and $\frac{1}{6}$ is $\frac{1}{3}$. This reasoning holds for every equality of the table.

   That this extension applies to all the succeeding tables can be seen from the following:

GENERATOR (1, 3)

$$4 + 12 = 3$$
$$8 + 24 = 6$$
$$12 + 36 = 9$$
$$\cdots,$$

GENERATOR $(1, 4)$

$$3 - 12 = 4$$
$$6 - 24 = 8$$
$$9 - 36 = 12$$
$$\cdots,$$

GENERATOR $(1, 4)$

$$5 + 20 = 4$$
$$10 + 40 = 8$$
$$15 + 60 = 12$$
$$\cdots,$$

GENERATOR $(1, 5)$

$$4 - 20 = 5$$
$$8 - 40 = 10$$
$$12 - 60 = 15$$
$$\cdots.$$

Thus, by an examination of simple equalities, we are led to:

Further extension of the G rule: If one of two unit fractions is K times the other, then their difference is found by dividing $(K - 1)$ into the larger number, providing the answer is an integer.

And again it can be said that, whether or not the ancient Egyptians used this rule or not, anyone working today in Egyptian mathematics will find it immeasurably useful in checking computations.

6 THE RECTO OF THE RHIND MATHEMATICAL PAPYRUS

THE DIVISION OF 2 BY THE ODD NUMBERS 3 TO 101

The Recto of the RMP occupies almost the first third of the roll, whose overall length is 18 feet and whose height is 13 inches. It was copied by the scribe Aᶜh-mosè during the period of the Hyksos or Shepherd Kings (about 1650 B.C.) from writings made about 200 years earlier. The papyrus is written in hieratic characters, and reads from right to left. It contains some 87 mathematical problems. These are preceded by a table of the division of 2 by the odd numbers 3 to 101, the answers being expressed as the sums of unit fractions (Figure 6.1).

For his introduction the scribe writes (Chace's translation) :*
Accurate reckoning. The entrance into the knowledge of all existing things and all obscure secrets. This book was copied in the year 33, in the 4th month of the inundation season, under the majesty of the king of Upper and Lower Egypt, ᶜA-user-Rêᶜ,† endowed with life, in likeness to writings of old made in the time of the king of Upper and Lower Egypt, Ne-ma-ᶜet-Rêᶜ.‡ It is the scribe Aᶜh-mosè who copies this writing.

This Recto Table (Table 6.1) is the most extensive and complete of all the arithmetical tables to be found in the Egyptian papyri that have come down to us. It must have been one of the most useful of all the scribes' reference tables, and would have been regarded, we might say, in about the same light as a modern mathematician would regard a set of logarithm tables.

That this comparison is not an overstatement is evidenced by the fact that many variations and extensions of the table have been found on much later papyri, ostraca, and wooden and other tablets, dating

---

* RMP. A. B. Chace; L. Bull; H. P. Manning; and R. C. Archibald, *The Rhind Mathematical Papyrus*, Mathematical Association of America, Oberlin, Ohio, 1927 (Vol. 1), 1929 (Vol. 2). See Vol. 1, p. 49.

† Probably ᶜAweserrēᶜ Apōpi, the sixth Hyksos ruler. See Gardiner, *Egypt of the Pharaohs*, Oxford University Press, London, 1964, p. 158 and p. 443.

‡ Nemaᶜrēᶜ Ammenemēs III of the Twelfth Dynasty, approximate dates 1842–1797, according to Gardiner, *Egypt of the Pharaohs*. Chace gives 1849–1801, following Breasted, *History of Egypt*, New York, 1911 edition.

FIGURE 6.1

The beginning of the Recto of the RMP. Shown here are the divisions $2 \div 3, 5, \ldots, 27$, preceded by the scribe's introductory remarks. The portion shown is 20 inches long. Courtesy British Museum.

from up to as many as 2,000 years after it was first written. We find, for example, unit fraction equalities for the numbers 2, 3,..., 101, each divided by 2, 3,..., 101, as late as the sixth century A.D. However more sophisticated and advanced the mathematics of the Greeks, Romans, Arabs, and Byzantines may have been, not one of these nations over this long period of time had been able to devise a more efficient technique for dealing with the simple common fraction p/q.

Thus the ancient Egyptian scribe, being required to divide 9 loaves equally among 10 men (RMP 6), worked it out that each man would receive $\bar{3}\ \bar{5}\ \overline{30}$ of a loaf. Again, in RMP 66, 3,200 ro* of fat are issued for a year, and it is calculated by the scribe that this is equivalent to using 8 $\bar{3}\ \overline{10}\ \overline{2190}$ ro per day. These unit fractions may appear clumsy, yet 2,200 years later a Greek papyrus† shows us—in a quite comprehensive table—that one-seventeenth of a silver talent (equal to 6,000 copper drachmas) is equal to 352 $\bar{2}\ \bar{3}\ \overline{17}\ \overline{34}\ \overline{51}$ drachmas. Thus we see that the Greeks in their own arithmetical notation‡ retained the ancient Egyptian unit fractions. Indeed, the papyrus material upon which the Greek table was inscribed also came from Egypt, as probably did the brush and ink as well.

And now in the twentieth century A.D., nearly 4,000 years after the Egyptians first devised their system for fractions, modern mathematicians have tried to determine what principles and processes the ancient Egyptian scribes used in preparing the Recto table. How was it possible for them, with only a knowledge of the twice-times table and an ability to find two-thirds of any integral or fractional number, to calculate unit fractional equivalents of $\frac{2}{5}$, $\frac{2}{7}$, $\frac{2}{9}$,..., $\frac{2}{101}$ without a single arithmetical error? And how did it come about that, of all the many thousands of possible answers to these decompositions, those recorded by the scribe of the RMP were in almost every case the simplest and best possible, by his own prescribed standards?

Some of the mathematicians of the late nineteenth and the twentieth centuries who have discussed these Egyptian unit fractions are

---

* A unit of measure. See Chapter 20.
† Michigan Papyrus 146. University of Michigan, Ann Arbor.
‡ The Greeks used the letters of their alphabet: $\alpha = 1, \beta = 2, \gamma = 3, \ldots$.

Eisenlohr (1877), Favarro (1879), Sylvester (1880), Collignon (1881), Shack-Schackenburg (1882), Tannery (1884), Mansion (1888), Bobynin (1890), Loria (1892), Hultsch (1895), Simon (1907), Griffith (1911), Vetter (1922), Vasconcellos (1923), Peet (1923), Neugebauer (1926), Gillain (1927), Chace (1927), Vogel (1929), Van der Waerden (1937), Hogben (1945), Struik (1948), Becker (1951), Bruins (1952), and Vogel (1958). Here are some comments of these writers.

. . . the very beautiful ancient Egyptian method of expressing all fractions under the form of a sum of the reciprocals of continually increasing integers.   SYLVESTER, 1882

Les décompositions sont toujours, à un point de vue ou à un autre, plus simple que toute autre décomposition possible. (The decompositions are always, from one point of view or another, simpler than any other decompositions.)   MANSION, 1888

Attempts to explain it [i.e., the method of the Egyptians] . . . have hitherto not succeeded.   HULTSCH, 1895

The men who designed the pyramids must have had insight into scientific principles, hardly credited to the Egyptians from their written documents alone.   GRIFFITH, 1911

Of the discussions which I have seen, the clearest is that by Loria. But no formula or rule has been discovered that will give all the results of the table, and Loria expressly says that he does not attempt to indicate how the old Egyptians obtained them.   CHACE, 1927

The Recto is a monument to the lack of scientific attitude of mind. PEET, 1931

They went to extraordinary pains to split up fractions like $2/43$ into a sum of unit fractions . . . a procedure as useless as it was ambiguous. The Greeks and Alexandrians continued this extraordinary performance.   HOGBEN, 1945

All available texts point to an Egyptian mathematics of rather primitive standards.   STRUIK, 1948

Das prinzip der Berechnung scheint kein einheitliches zu sein.   (The principle of calculation does not seem to be uniform.)   BECKER, 1951

One would hardly have expected such diverse and contradictory opinions from among such competent critics.

The usual statement made by students of the Recto decompositions is that the scribe was seeking the "simplest" value available; but just what constituted a "simplest" equality is seldom made quite clear. Opinions are quite varied on the precepts, standards, or tests by which the scribe was guided in his choice of values from the hundreds available to him. Some previous investigators have attempted to give the scribe's possible precepts. I here present the five precepts which I believe were the scribe's primary guide. The fifth precept has not been suggested, to my knowledge, by any previous writer.

CANON FOR THE RECTO OF THE RMP

PRECEPT 1

Of the possible equalities, those with the smaller numbers are preferred, but *none* as large as 1,000.*

PRECEPT 2

An equality of only 2 terms is preferred to one of 3 terms, and one of 3 terms to one of 4 terms, but an equality of more than 4 terms is *never* used.

PRECEPT 3

The unit fractions are always set down in descending order of magnitude, that is, the smaller numbers come first, but *never* the same fraction twice.

PRECEPT 4

The smallness of the first number is the main consideration, but the scribe will accept a *slightly* larger first number, if it will *greatly* reduce the last number.

PRECEPT 5

Even numbers are preferred to *odd* numbers,† even though they might be larger, and even though the numbers of terms might thereby be increased.

* The largest number in the Recto is 890. Of the 128 numbers of the table only 11 exceed 500.
† There are 104 even and only 24 odd numbers used in the table.

TABLE 6.1
The RMP Recto Table: Two divided by 3, 5, 7, . . . , 101. Note that all unit fractions are here written without overbars.

Divisor	Unit Fractions				Divisor	Unit Fractions			
3	$\overline{3}$				53	30	318	795	
5	3	15			55	30	330		
7	4	28			57	38	114		
9	6	18			59	36	236	531	
11	6	66			61	40	244	488	610
13	8	52	104		63	42	126		
15	10	30			65	39	195		
17	12	51	68		67	40	335	536	
19	12	76	114		69	46	138		
21	14	42			71	40	568	710	
23	12	276			73	60	219	292	365
25	15	75			75	50	150		
27	18	54			77	44	308		
29	24	58	174	232	79	60	237	316	790
31	20	124	155		81	54	162		
33	22	66			83	60	332	415	498
35	30	42			85	51	255		
37	24	111	296		87	58	174		
39	26	78			89	60	356	534	890
41	24	246	328		91	70	130		
43	42	86	129	301	93	62	186		
45	30	90			95	60	380	570	
47	30	141	470		97	56	679	776	
49	28	196			99	66	198		
51	34	102			101	101	202	303	606

In 1967 an electronic computer was programmed to calculate all the possible unit-fraction expressions of each of the divisions of 2 by the odd numbers 3, 5, 7, . . . , 101, in order to compare the decompositions given by the scribe of the RMP with the thousands of possible forms.* Such a comparison between the calculations of an ancient

* Professor C. L. Hamblin of the School of Philosophy, University of New South Wales, kindly programmed the KDF-9 computer of Sydney University to do this, and supervised the production of the results. The time taken by the computer was 5 hours.

Egyptian scribe and the 22,295 values produced by a twentieth-century computer, separated by a time of nearly 4,000 years, will undoubtedly be of great interest to historians of mathematics.

To illustrate the application of the five precepts of the canon to the divisions of the Recto Table, let us take the division of 2 by 45. The computer lists 1,967 possible decompositions of $2 \div 45$ into sums of not more than four unit fractions, of which 7 have 2 terms, 134 have 3 terms, and 1,826 have 4 terms. Since there are seven 2-term decompositions, it would seem that it would be unnecessary to seek among the 1,826 4-term values for the "simplest" value, and possibly also the 3-term values may not be needed. Indeed, none of these 3- or 4-term values may even have come to the scribe's attention. The seven 2-term equalities given by the computer are:

A	$\overline{24}$	$\overline{360}$
B	$\overline{25}$	$\overline{225}$
C	$\overline{27}$	$\overline{135}$
D	$\overline{30}$	$\overline{90}$
E	$\overline{35}$	$\overline{63}$
F	$\overline{36}$	$\overline{60}$
G	$\overline{45}$	$\overline{45}.$

Following the canon we at once rule out G, since the fraction $\overline{45}$ is repeated. Next, E, C, and B are eliminated in that order, because $35 > 27 > 25$, and also because they contain only odd numbers (Precepts 1 and 5). Then there remain A, D, and F. Of these, the first to go is A, for although 24 is less than 30 and 36, the 3-digit number 360 is so much greater than 90 and 60, that Precept 4 forbids the choice. The choice then lies between D and F, and here Precept 4 is of little help, because 60 is not sufficiently less than 90 to outweigh the excess of 36 over 30. In decisions like this, the scribe's judgments appear to have been flexible; further, we note that other considerations favor D and not F, because

$$2 \div 45 = \overline{30} \quad \overline{90},$$

121876

is easily derived by multiplication of the earlier values

$$2 \div 15 = \overline{10} \quad \overline{30},$$
$$2 \div 9 = \overline{6} \quad \overline{18}$$

by 3 and 5, respectively.

There may have been perhaps a further reason for the scribe's choice of D, which would apply to all divisors which are multiples of 3. As we have seen (RMP 61B, p. 35), the scribe's rule for finding two-thirds of any odd fraction can be briefly stated as "twice it, and six times it." Then, since

$$2 \div 45 \equiv \overline{\overline{3}} \text{ of } \overline{15}$$
$$= \overline{30} \quad \overline{90},$$

the scribe would enter this decomposition into the Recto Table. We can conclude that, in this division as elsewhere, the computer did not find a decomposition superior to that given by the ancient scribe.

CONCERNING PRIMES

Of the 50 divisors of 2 used in the RMP Recto Table, 25 are prime numbers, and these are the divisors which must have given the scribe most food for thought. An analysis of the computer output shows that 2-term decompositions are rare for primes, and that there are very few 3-term values; so that the scribe's search for the simplest unit-fraction equalities must have presented considerable difficulties. Thus there are *no 2-term or 3-term* decompositions for the prime divisors 61, 79, 83, 89, and 101; and there are *no 2-term* decompositions for the prime divisors 47, 53, 59, 67, 71, 73, and 97. There remain 13 prime divisors for which, according to the computer, *only one 2-term* decomposition is possible; all except four of these are unsuitable by the precepts given on p. 49.

For 2 divided by 5, 7, 11, and 23, the scribe accepts the only 2-term equalities available, but for 2 divided by 13, 17, 19, 29, 31, 37, 41, and 43 the solitary 2-term equalities are not acceptable, and he has to go further afield. Table 6.2 shows this. The asterisks show the group to

TABLE 6.2
Prime divisors between 3 and 101. Asterisks show the
group out of which the scribe chose his "simplest"
decomposition for each divisor.

Prime divisor	Number of 2-term values	Number of 3-term values	Number of 4-term values
3	1	4	48
5	1*	8	260
7	1*	13	306
11	1*	16	367
13	1	12*	423
17	1	11*	467
19	1	16*	256
23	1*	18	368
29	1	8	203*
31	1	8*	155
37	1	6*	90
41	1	7*	179
43	1	6	117*
47	0	2*	54
53	0	1*	23
59	0	1*	19
61	0	0	7*
67	0	2*	12
71	0	2*	23
73	0	1	9*
79	0	0	3*
83	0	0	3*
89	0	0	6*
97	0	1*	7
101	0	0	1*

which the scribe needed to go to find what he regarded as the simplest
or most suitable decomposition. For $2 \div 3$, the scribe merely put $\bar{3}$,
even though he certainly knew he could have written $\bar{2}\ \bar{6}$.

FURTHER COMPARISON OF THE SCRIBE'S AND THE COMPUTER'S DECOMPOSITIONS

$2 \div 5$ The scribe's decomposition is $\bar{3}\ \overline{15}$, and from the 269 expressions computed by KDF-9 we find that this is the *solitary* 2-term value

possible. Although both numbers are odd, they are so small that they do not induce a preference for any 3-term equality which may contain only even numbers. The eight 3-term decompositions recorded by KDF-9 are

A	3	$\overline{16}$	$\overline{240}$
B	3	$\overline{18}$	$\overline{90}$
C	3	$\overline{20}$	$\overline{60}$
D	3	$\overline{24}$	$\overline{40}$
E	$\overline{4}$	$\overline{7}$	$\overline{140}$
F	$\overline{4}$	$\overline{8}$	$\overline{40}$
G	$\overline{4}$	$\overline{10}$	$\overline{20}$
H	$\overline{4}$	$\overline{12}$	$\overline{15}$.

F and G are the only ones having all numbers even, but he preferred $\overline{3}\ \overline{15}$ because Precepts 1 and 2, taken together, outweigh Precept 5.

$2 \div 7$ The scribe wrote $\overline{4}\ \overline{28}$; the computer shows that of the 320 possible decompositions this is the only one consisting of only two terms, and they are both even numbers. The scribe has therefore chosen the simplest and best-possible value by his standards.

$2 \div 9$ For this division, the computer lists 516 possible decompositions, of which the only ones with two terms are $\overline{5}\ \overline{45}$ and $\overline{6}\ \overline{18}$. Since the latter has even numbers and 18 is less than 45, Precepts 1, 2, and 5 decide that this is the simplest value. Furthermore, $\overline{6}\ \overline{18}$ is derivable from

$$2 \div 3 = \overline{2}\ \ \overline{6},$$

on multiplying through by 3.

   Indeed, all the subsequent divisors that are multiples of 3 are derivable by the successive multiplication through by 3, 5, 7, . . ., 33, so that they are all 2-term values that conform to the canon. However, we note that KDF-9 found other 2-term decompositions for these multiples of 3. We therefore list these other decompositions (Table 6.3), and compare them with those written down by the scribe. Even if every equality in Table 6.3 had been known to the scribe, it would still appear that nowhere has he made an obviously bad choice. In one or two instances we might have some slight doubts, but even for these Precept 4 vindicates the scribe's choice, the smaller first number deciding for him. Thus,

for	he chose		he rejected	
$2 \div 15$	$\overline{10}$	$\overline{30}$	$\overline{12}$	$\overline{20}$
$2 \div 45$	$\overline{30}$	$\overline{90}$	$\overline{36}$	$\overline{60}$
$2 \div 63$	$\overline{42}$	$\overline{126}$	$\overline{56}$	$\overline{72}$
$2 \div 75$	$\overline{50}$	$\overline{150}$	$\overline{60}$	$\overline{100}$
$2 \div 99$	$\overline{66}$	$\overline{198}$	$\overline{90}$	$\overline{110}$

$2 \div 11$ Of the 384 decompositions given by the computer, only one 2-term expression occurs, $\overline{6}\ \overline{66}$, which the scribe recorded. This is clearly the simplest possible; both numbers are even.

$2 \div 13$ Again there was only one 2-term decomposition, $\overline{7}\ \overline{91}$, among the 436 listed by the computer. But the scribe would not accept it, both numbers being odd (Precept 5). Now there are twelve 3-term expressions available, but only four of them have three even numbers. These are

A	$\overline{8}$	$\overline{36}$	$\overline{936}$
B	$\overline{8}$	$\overline{40}$	$\overline{260}$
C	$\overline{8}$	$\overline{52}$	$\overline{104}$
D	$\overline{10}$	$\overline{20}$	$\overline{260}.$

Of these he chose C, by virtue of Precept 1, and this is clearly the best choice, noting the 104.

$2 \div 15$ See $2 \div 9$.

$2 \div 17$ As was the case for 11 and 13, only one 2-term decomposition is possible, namely $\overline{9}\ \overline{153}$, out of the 479 given by the computer. However, the scribe rejected it. Both numbers are odd, and perhaps we can find a 3-term equality with numbers less than 153. The eleven

TABLE 6.3

Two-term decompositions of 2 divided by multiples of 3. Decompositions that were chosen by the scribe are noted by an asterisk.

Divisor	Total number of computer values	Number of 2-term values	Computer values	
15	1158	4	$\bar{8}$	$\overline{120}$
			$\bar{9}$	$\overline{45}$
			$\overline{10}$	$\overline{30}$*
			$\overline{12}$	$\overline{20}$
21	1190	4	$\overline{11}$	$\overline{231}$
			$\overline{12}$	$\overline{84}$
			$\overline{14}$	$\overline{42}$*
			$\overline{15}$	$\overline{35}$
27	733	3	$\overline{14}$	$\overline{378}$
			$\overline{15}$	$\overline{135}$
			$\overline{18}$	$\overline{54}$*
33	1016	4	$\overline{17}$	$\overline{561}$
			$\overline{18}$	$\overline{198}$
			$\overline{21}$	$\overline{77}$
			$\overline{22}$	$\overline{66}$*
39	894	4	$\overline{20}$	$\overline{780}$
			$\overline{21}$	$\overline{273}$
			$\overline{24}$	$\overline{104}$
			$\overline{26}$	$\overline{78}$*
45	1967	6	$\overline{24}$	$\overline{360}$
			$\overline{25}$	$\overline{225}$
			$\overline{27}$	$\overline{135}$
			$\overline{30}$	$\overline{90}$*
			$\overline{35}$	$\overline{63}$
			$\overline{36}$	$\overline{60}$
51	595	3	$\overline{27}$	$\overline{459}$
			$\overline{30}$	$\overline{170}$
			$\overline{34}$	$\overline{102}$*
57	645	3	30	570
			33	209
			38	114*
63	1607	6	$\overline{33}$	$\overline{693}$
			$\overline{35}$	$\overline{315}$
			$\overline{36}$	$\overline{252}$

TABLE 6.3 (*continued*)

Divisor	Total number of computer values	Number of 2-term values	Computer values	
63	1067	6	$\overline{42}$	$\overline{126}$*
			$\overline{45}$	$\overline{105}$
			$\overline{56}$	$\overline{72}$
69	500	3	$\overline{36}$	$\overline{828}$
			$\overline{39}$	$\overline{299}$
			$\overline{46}$	$\overline{138}$*
75	884	6	$\overline{39}$	$\overline{975}$
			$\overline{40}$	$\overline{600}$
			$\overline{42}$	$\overline{350}$
			$\overline{45}$	$\overline{225}$
			$\overline{50}$	$\overline{150}$*
			$\overline{60}$	$\overline{100}$
81	339	2	$\overline{45}$	$\overline{405}$
			$\overline{54}$	$\overline{162}$*
87	102	2	$\overline{48}$	$\overline{464}$
			$\overline{58}$	$\overline{174}$*
93	58	2	$\overline{51}$	$\overline{527}$
			$\overline{62}$	$\overline{186}$*
99	710	5	$\overline{54}$	$\overline{594}$
			$\overline{55}$	$\overline{495}$
			$\overline{63}$	$\overline{231}$
			$\overline{66}$	$\overline{198}$*
			$\overline{90}$	$\overline{110}$

3-term values are

A	$\overline{9}$	$\overline{204}$	$\overline{612}$
B	$\overline{9}$	$\overline{234}$	$\overline{442}$
C	$\overline{10}$	$\overline{65}$	$\overline{442}$
D	$\overline{10}$	$\overline{68}$	$\overline{340}$
E	$\overline{10}$	$\overline{85}$	$\overline{170}$
F	$\overline{10}$	$\overline{90}$	$\overline{153}$
G	$\overline{12}$	$\overline{34}$	$\overline{204}$
H	$\overline{12}$	$\overline{36}$	$\overline{153}$
I	$\overline{12}$	$\overline{51}$	$\overline{68}$
J	$\overline{13}$	$\overline{26}$	$\overline{442}$
K	$\overline{17}$	$\overline{18}$	$\overline{306}$.

Of these, D and G are the only ones composed wholly of even numbers, and if Precepts 4 and 5 are considered, one would expect the choice to fall on G. But it does not! Precept 1 on smaller numbers must have prevailed, for the scribe selected I, even though one of the numbers is odd. We cannot know whether the scribe was aware of all these possibilities, but whether he was or not, he certainly found the only 3-term decomposition consisting of two-digit numbers. Even if he had looked among the 467 4-term decompositions, he would have found only 3 consisting of two-digit numbers, but all with numbers much greater than $\overline{12}\ \overline{51}\ \overline{68}$; so that, however he did it, we can only compliment him on an amazingly successful search.

$2 \div 19$ The only 2-term decomposition here is $\overline{10}\ \overline{190}$, and the scribe must have thought hard before he rejected it. Of the sixteen 3-term values available, only five consist wholly of even numbers. These are

A	$\overline{10}$	$\overline{240}$	$\overline{912}$
B	$\overline{12}$	$\overline{48}$	$\overline{912}$
C	$\overline{12}$	$\overline{60}$	$\overline{190}$
D	$\overline{12}$	$\overline{76}$	$\overline{114}$
E	$\overline{16}$	$\overline{24}$	$\overline{912}$.

Precept 1 must have prevailed, for the scribe chose D, as having the smallest numbers, for although 10 is less than 12, 114 is much less than 190 (see Precept 4).

$2 \div 21$ See $2 \div 9$.

$2 \div 23$ Another prime number divisor, and consequently only one 2-term decomposition is possible ($\overline{12}\ \overline{276}$), which the scribe at once accepted. All of the eighteen 3-term decompositions have last terms much greater than 276, including the seven that contain only even numbers, so that whether he looked further afield or not, he chose here the simplest expression.

$2 \div 25$ The computer records 619 values, of which

A	$\overline{13}$	$\overline{325}$
B	$\overline{15}$	$\overline{75}$,

are the only 2-term expressions, from which the scribe naturally chose B, primarily no doubt because it is at once obtainable from $2 \div 5 = \overline{3}\ \overline{15}$, upon multiplication by 5. If the scribe had sought among the 28 possible 3-term decompositions for even numbers, he would have found only 7, all with much greater numbers,

A	$\overline{14}$	$\overline{140}$	$\overline{700}$
B	$\overline{14}$	$\overline{200}$	$\overline{280}$
C	$\overline{16}$	$\overline{80}$	$\overline{210}$
D	$\overline{20}$	$\overline{36}$	$\overline{450}$
E	$\overline{20}$	$\overline{40}$	$\overline{200}$
F	$\overline{20}$	$\overline{50}$	$\overline{100}$
G	$\overline{24}$	$\overline{30}$	$\overline{200}.$

However, if he had been aware of these 3-term expressions, the decomposition F must certainly have tempted him, but 15 and 75 are less than 20 and 100.

$2 \div 27$ See $2 \div 9$.

$2 \div 29$ Of the 212 decompositions given by the computer, we find 1 containing 2 terms, 8 containing 3 terms, and 203 containing 4 terms. The solitary 2-term decomposition is $\overline{15}\ \overline{435}$, where both terms are odd and 435 is fairly large, so the scribe looked further afield. The 3-term values are

A	$\overline{16}$	$\overline{232}$	$\overline{464}$
B	$\overline{16}$	$\overline{240}$	$\overline{435}$
C	$\overline{18}$	$\overline{87}$	$\overline{522}$
D	$\overline{18}$	$\overline{90}$	$\overline{435}$
E	$\overline{20}$	$\overline{58}$	$\overline{580}$
F	$\overline{20}$	$\overline{60}$	$\overline{435}$
G	$\overline{24}$	$\overline{40}$	$\overline{435}$
H	$\overline{29}$	$\overline{30}$	$\overline{870}.$

But he would have none of these either. All have numbers as large as 435 or larger, and all except A and E contain odd numbers. Perhaps he could do better among the 4-term equalities! Whether he knew of all these possibilities we cannot of course tell, but we do know that by the standards of the canon he found the very best available. For of the 203 decompositions listed by KDF-9, only three contain numbers less than 300. They are

$$
\begin{array}{ccccc}
\text{A} & \overline{24} & \overline{58} & \overline{174} & \overline{232} \\
\text{B} & \overline{29} & \overline{42} & \overline{174} & \overline{203} \\
\text{C} & \overline{29} & \overline{58} & \overline{87} & \overline{174}.
\end{array}
$$

By all the precepts of the canon he must choose A, and this is indeed the equality recorded in the Recto. We cannot but admire the skill with which the scribe found the equality with the smallest even numbers from the 212 available, and wonder just how he did it.

$2 \div 31$ Again a prime divisor with only one 2-term value, $\overline{16}\ \overline{496}$,* and although both terms are even numbers, 496 is much too large and perhaps the scribe can do better. Of the 163 other possibilities, 8 only have 3 terms, and from these he was able to locate $\overline{20}\ \overline{124}\ \overline{155}$, which has easily the smallest numbers, although he had to accept the odd number 155. Those decompositions with all even numbers are

$$
\begin{array}{cccc}
\text{A} & \overline{18} & \overline{144} & \overline{496} \\
\text{B} & \overline{20} & \overline{80} & \overline{496} \\
\text{C} & \overline{24} & \overline{48} & \overline{496},
\end{array}
$$

which we would expect him to have rejected, for he had already rejected the simpler $\overline{16}\ \overline{496}$ because of the large 496.

$2 \div 35$ This division by 35 is the only division in the whole of the Recto in which the scribe gives us any inkling of his method.† For $2 \div 35$, the computer lists 1,458 possible unit-fraction decompositions, of which only four contain two terms. These are

$$
\begin{array}{ccc}
\text{A} & \overline{18} & \overline{630} \\
\text{B} & \overline{20} & \overline{140} \\
\text{C} & \overline{21} & \overline{105} \\
\text{D} & \overline{30} & \overline{42}.
\end{array}
$$

* In a table of fractions dating from Greek times, published by Crum, *Coptic Ostraca*, London, 1902, p. 46, later discussed by Sethe, *Von Zahlen und Zahlworten bei den alten Ägyptern*, Trübner, Strassburg, 1916, occurs $2 \div 31 = \overline{31}\ \overline{62}\ \overline{93}\ \overline{186}$. Compare $2 \div 101$, RMP Recto.
† The method will be discussed in detail in Chapter 7.

Both on the score of smaller numbers and of even numbers, his choice should have fallen on D, and this is indeed the equality given in the Recto. There would have been no need to look at any of the other 1,454 values, nor to have considered deriving B from $(2 \div 7) \times 5 = (\overline{4}\ \overline{28}) \times 5 = (\overline{20}\ \overline{140})$, or C from $(2 \div 5) \times 7 = (\overline{3}\ \overline{15}) \times 7 = (\overline{21}\ \overline{105})$, although, of course, he may have checked on these. Clearly, of the 1,458 possibilities, the scribe chose the simplest.

$2 \div 37$ For this prime divisor only 97 answers are possible. The 2-term decomposition is $\overline{19}\ \overline{703}$, which is clearly unsuitable, because both numbers are odd, and 703 is far too large. Of the six 3-term answers, there is only one having all numbers even, $\overline{20}\ \overline{370}\ \overline{740}$, but these numbers are even larger than before, so it was also rejected. From among the remaining five, the scribe looked for the equality with the smallest numbers, and so selected $\overline{24}\ \overline{111}\ \overline{296}$, being forced to accept one odd number; but this is consistent with his usual procedure.

$2 \div 39$ See $2 \div 9$.

$2 \div 41$ In this division by a prime number the scribe really shows his capabilities in handling unit fractions. The sole 2-term decomposition is $\overline{21}\ \overline{861}$, which he immediately rejects. There remain 186 other decompositions, of which only 7 consist of 3 terms. Now four of these have $\overline{861}$ as their highest term, one has $\overline{902}$ as its highest term, and another has $\overline{984}$, all large numbers, and in addition, other terms are odd. The sole remaining 3-term expression is $\overline{24}\ \overline{246}\ \overline{328}$, by far the simplest of the lot, for all the 4-term decompositions listed by KDF-9 contain very much larger numbers. From the 187 decompositions available, the scribe certainly located that one which by his standards clearly is the best offering, and he is to be complimented on his success, whether accidental or not.

$2 \div 43$ The scribe's effort here is truly amazing. From the 124 decompositions listed by the computer, those having two or three terms are

A	$\overline{22}$	$\overline{946}$	
B	$\overline{23}$	$\overline{506}$	$\overline{946}$
C	$\overline{24}$	$\overline{264}$	$\overline{946}$
D	$\overline{24}$	$\overline{344}$	$\overline{516}$
E	$\overline{26}$	$\overline{143}$	$\overline{946}$
F	$\overline{30}$	$\overline{86}$	$\overline{645}$
G	$\overline{33}$	$\overline{66}$	$\overline{946}.$

Now B, E, F, and G contain odd numbers, and so are not acceptable, and although A, C, and D have only even numbers, they are far too large for the scribe's purposes, even though A contains only two terms. To seek smaller numbers, the scribe needed to search among the 117 4-term equalities, and he would have had to accept odd numbers. Looking through this group, we find that 83 of them have their fourth term greater than 900, three of them have their fourth term greater than 800, one of them greater than 700, eighteen of them greater than 600, and eleven of them greater than 500. This is a total of 116, so that one only remains. This sole remaining 4-term equality has its fourth term 301, and although this is an odd number, we find that $\overline{42}\ \overline{86}$ $\overline{129}\ \overline{301}$ has the smallest numbers of any of the possible values. The smallest high number contained in any of the other 123 possible answers is 516, whether they be 2-term, 3-term, or 4-term equalities, and one can only remain lost in hopeless admiration of the ancient Egyptian scribe, who could, with the meager arithmetical tools at his disposal, so unerringly locate this value among the 124 available.

$2 \div 45$ See $2 \div 9$ and pages 51–52.

$2 \div 47$ Here no 2-term decompositions are possible, and only two 3-term equalities occur in the computer list, which totals 56 entries. The 3-term decompositions are

A	$\overline{28}$	$\overline{188}$	$\overline{658}$
B	$\overline{30}$	$\overline{141}$	$\overline{470},$

from which the scribe selected B because of the much smaller 470, despite 141 being odd. However, he may have had some doubts here, for there is little to choose between them. Precept 4 no doubt decided for him.

2 ÷ 49   Of the 371 possible decompositions, the sole 2-term one is $\overline{28}\ \overline{196}$, with both numbers even and not too large. This is the value given by the scribe, and it could have been obtained from $2 \div 7 = \overline{4}\ \overline{28}$ multiplied by 7. There are thirty 2-term values possible, all with very high numbers, but probably none of them even came to his attention. His choice here was clearly the best available.

2 ÷ 51   See 2 ÷ 9.

2 ÷ 53   The computer located only 24 possible values for 2 ÷ 53, none at all consisting of two terms, and only one, $\overline{30}\ \overline{318}\ \overline{795}$, of three terms, and this is the value the scribe chose for the Recto. There was certainly not much to choose from in this division, for of the twenty-three 4-term equalities, all contain very high odd numbers except $\overline{42}\ \overline{106}\ \overline{318}\ \overline{742}$, which he rejected by virtue of Precept 4, but which he might have accepted by virtue of Precept 5. If this equality did come to his attention, he must have thought very deeply before deciding.

2 ÷ 55   There are 1,128 decompositions of 2 ÷ 55, of which 1,052 have 4 terms, 73 have 3 terms, and only 3 have 2 terms. The numbers in this last group are so much smaller than all the others that the scribe would not have fared any better if he had looked elsewhere, and so his choice was reduced to

A	$\overline{30}$	$\overline{330}$
B	$\overline{33}$	$\overline{165}$
C	$\overline{40}$	$\overline{88}$.

Choice B would have been rejected because of the odd numbers, which are inevitable, as it is derived from $2 \div 5 = \overline{3}\ \overline{15}$ multiplied by 11. However, A comes just as easily from $2 \div 11 = \overline{6}\ \overline{66}$ multiplied by 5, and it contains only even numbers. If C did come to the scribe's attention, Precept 4 may have swayed him in favor of A, for 30 being less than 40 is his main consideration, even though 330 is much greater than 88. We can imagine some queries in the scribe's mind, but the easy derivation from 2 ÷ 11 no doubt persuaded him to choose A.

2 ÷ 57   See 2 ÷ 9.

$2 \div 59$ Here the scribe and the computer at once concur. Of the 20 possibilities, 19 contain 4 terms, not one of which is composed wholly of even numbers. Indeed they average two odd numbers each, and they are all large numbers. The sole 3-term decomposition is $\overline{36}\ \overline{236}\ \overline{531}$, which, although it has one odd number in it, is still the simplest expression available, no other equality having its first term less than 36 nor its last term less than 531. No 2-term equality is possible for $2 \div 59$. This is the decomposition the scribe records in the Recto.

$2 \div 61$ Agreement between the scribe and the KDF-9 is even more obvious here. There are only 7 possibilities, all of which contain 4 terms, and only 2 of these consist wholly of even numbers. They are

$$
\begin{array}{lcccc}
A & \overline{40} & \overline{244} & \overline{488} & \overline{610} \\
B & \overline{48} & \overline{122} & \overline{366} & 976.
\end{array}
$$

Since in A both the first and last terms are less than in B, this is the value chosen by the scribe. It is clearly the best available.

$2 \div 63$ See $2 \div 9$.

$2 \div 65$ There are 865 decompositions listed by the computer, but only 3 of these contain 2 terms. They are

$$
\begin{array}{lcc}
A & \overline{35} & \overline{455} \\
B & \overline{39} & \overline{195} \\
C & \overline{45} & \overline{117}.
\end{array}
$$

All numbers are odd, so none has preference on that score. But B is obtainable directly from $2 \div 5 = \overline{3}\ \overline{15}$ on multiplication by 13, and this probably weighed with the scribe when comparing C, considering Precept 4 of the canon. Had he looked for 3- or 4-term values he would have fared no better; the best offering is $2 \div 65 = \overline{60}\ \overline{130}\ \overline{156}$.

$2 \div 67$ No 2-term decomposition exists out of a total of twenty-one, and there are only two 3-term expressions, which are

$$
\begin{array}{lccc}
A & \overline{40} & \overline{335} & \overline{536} \\
B & \overline{42} & \overline{201} & 938.
\end{array}
$$

Both contain one odd number; but in A both the first and last terms are smaller, and so the scribe chooses A. Had he looked at the 4-term expressions, he would have found only one without odd numbers, $\overline{42}\ \overline{268}\ \overline{804}\ \overline{938}$, of which the numbers are far too large to interest him. He thus chose the best possible decomposition.

$2 \div 69$ See $2 \div 9$.

$2 \div 71$ Of 25 possible equalities, all except two have 4 terms, and these are

$$
\begin{array}{cccc}
\text{A} & \overline{40} & \overline{568} & \overline{710} \\
\text{B} & \overline{42} & \overline{426} & \overline{497}.
\end{array}
$$

The scribe at once chose A, 40 being less than 42, and although 497 is less than 710, it is an odd number. Had the last term of B been even, he would no doubt have selected it by virtue of Precepts 4 and 5.

$2 \div 73$ The scribe had a poor selection to choose from here, for of the 10 possible decompositions, all except $\overline{44}\ \overline{292}\ \overline{803}$ have 4 terms with large numbers, 876 occurring twice, 803 five times, and 703 once. In addition they all contain odd numbers. The scribe chose the one remaining expression, $\overline{60}\ \overline{219}\ \overline{292}\ \overline{365}$, which is the one with the smallest numbers. By his standards he certainly found the best decomposition available.

$2 \div 75$ See $2 \div 9$.

$2 \div 77$ Of the 741 decompositions generated by the computer, only 3 have two terms:

$$
\begin{array}{ccc}
\text{A} & \overline{42} & \overline{462} \\
\text{B} & \overline{44} & \overline{308} \\
\text{C} & \overline{63} & \overline{99}.
\end{array}
$$

Choice C can be rejected because of its odd numbers, even though 99 is less than 308. Choice A is derivable from $2 \div 11 = \overline{6}\ \overline{66}$ on multiplying by 7, and B from $2 \div 7 = \overline{4}\ \overline{28}$ on multiplying by 11, and so in conformity with Precepts 2, 4, and 5, we, as did the scribe, select B, clearly the best choice available, without seeking further among the 738 other 3- and 4-term possibilities.

2 ÷ 79   The prime divisor 79 has the least number of decompositions of all divisions except 101, namely, 3:

A	$\overline{60}$	$\overline{237}$	$\overline{316}$	$\overline{790}$
B	$\overline{60}$	$\overline{158}$	$\overline{790}$	$\overline{948}$
C	$\overline{79}$	$\overline{99}$	$\overline{711}$	$\overline{869}$.

By Precept 1 the scribe should have selected A because of the smaller numbers, even though one of them is odd. This the scribe did; it is the best available choice.

2 ÷ 81   See 2 ÷ 9.

2 ÷ 83   Like 79, the divisor 83 permits of only 3 decompositions:

A	$\overline{56}$	$\overline{332}$	$\overline{581}$	$\overline{664}$
B	$\overline{60}$	$\overline{332}$	$\overline{415}$	$\overline{498}$
C	$\overline{60}$	$\overline{249}$	$\overline{415}$	$\overline{996}$.

All contain odd numbers, and the only test applicable here is Precept 1 on the smallness of numbers, and unerringly the scribe found and recorded choice B in the Recto Table.

2 ÷ 85   Of 290 possible values, 255 have 4 terms, 32 have 3 terms, and 3 have 2 terms. These last are

A	$\overline{45}$	$\overline{765}$
B	$\overline{51}$	$\overline{255}$
C	$\overline{55}$	$\overline{187}$.

All six numbers are odd, but A has a very large second term, so the scribe's choice lay between B and C, and Precept 4 points at once to B. Furthermore, B is easily derivable from $2 ÷ 5 = \overline{3}\ \overline{15}$ on multiplying by 17, so that the scribe had the simplest value in the Recto Table.

2 ÷ 87   See 2 ÷ 9.

2 ÷ 89   The computer records only 6 possible values:

A	$\overline{54}$	$\overline{594}$	$\overline{801}$	$\overline{979}$
B	$\overline{55}$	$\overline{495}$	$\overline{801}$	$\overline{979}$
C	$\overline{60}$	$\overline{356}$	$\overline{534}$	$\overline{890}$

D	$\overline{63}$	$\overline{231}$	$\overline{801}$	$\overline{979}$
E	$\overline{63}$	$\overline{267}$	$\overline{623}$	$\overline{801}$
F	$\overline{66}$	$\overline{198}$	$\overline{801}$	$\overline{979}$.

All have four terms, and all numbers are high. A, B, D, E, and F have respectively 2, 4, 4, 4, and 2 odd numbers, while the remaining decomposition C is the only one consisting wholly of even numbers, and this is indeed the expression that the scribe recorded in the Recto. The denominator 890 in the last term of this decomposition is the largest number occurring anywhere in the Recto Table.

$2 \div 91$ Since $91 = 7 \times 13$, and thus is not a prime, undoubtedly the scribe looked at the values derivable from his earlier divisions by 7 and 13. He would have found from $2 \div 7$, $(\overline{4}\ \overline{28}) \times 13 = (\overline{52}\ \overline{364})$, and from $2 \div 13$, $(\overline{7}\ \overline{91}) \times 7 = (\overline{49}\ \overline{637})$. Now if this was all he did, he would quite naturally have selected $\overline{52}\ \overline{364}$ according to the canon, and that would have been the end of the matter. But he looked further afield for something better, and he found it. The computer records 216 possibilities, of which 185 have 4 terms, 28 have 3 terms, and 3 have 2 terms. These last are

A	$\overline{49}$	$\overline{637}$
B	$\overline{52}$	$\overline{364}$
C	$\overline{70}$	$\overline{130}$.

We already know how he could have found A and B, but how did he find the only other 2-term equality C, in which 130 is less than half 364? By Precepts 1 and 4, he has certainly found the simplest value in C, however he did it,* for a careful examination of the 214 remaining 3- and 4-term values shows that in every case, the last term is more than double 130, and in most cases is 3, 4, 5 and even 6 times it. Whether the scribe even bothered to look at any 3-term or 4-term values we shall never know; what we do know is that if he had, he would have found nothing as good as the equality recorded in the Recto Table.

* We can show by algebra that $2 \div ab$ (where a and b are both odd, and $a + b = 20$) equals $\overline{10a} + \overline{10b}$. Then $2 \div 19$ is $\overline{10}\ \overline{190}$, $2 \div 51$ is $\overline{30}\ \overline{170}$, $2 \div 75$ is $\overline{50}\ \overline{150}$, $2 \div 91$ is $\overline{70}\ \overline{130}$, and $2 \div 99$ is $\overline{90}\ \overline{110}$.

$2 \div 93$ See $2 \div 9$.

$2 \div 95$ There are 148 values recorded by the computer, of which 116 have 4 terms, 29 have 3 terms, and 3 have 2 terms. These last are

$$
\begin{array}{ccc}
A & \overline{50} & \overline{950} \\
B & \overline{57} & \overline{285} \\
C & \overline{60} & \overline{228}.
\end{array}
$$

Now by the canon, C should have been his immediate choice here; for 950 in A is far too large a number, and 57 and 285 in B are both odd. But this is not the value the scribe gives. So far, we have been unable seriously to challenge the scribe's choice of values in the Recto Table; but here, perhaps, he faltered. "Even Homer nodded" on occasions. The equality the scribe records is

$$ \overline{60} \qquad \overline{380} \qquad \overline{570}, $$

a 3-term value which is in fact equivalent to

$$ \overline{60} \qquad \overline{228}, $$

for $\overline{380}\ \overline{570} = \overline{228}$, which he should have known.* What he must have done here was to note that $95 = 5 \times 19$, then looked for $2 \div 5$ multiplied by 19, and also $2 \div 19$ multiplied by 5; the same technique as we judge him to have adopted for $2 \div 91$. From $2 \div 5 = (\overline{3}\ \overline{15})$ he has, on multiplying by nineteen, $(\overline{57}\ \overline{285})$, *both odd numbers*. From $2 \div 19 = (\overline{12}\ \overline{76}\ \overline{114})$, he has, on multiplying by five, $(\overline{60}\ \overline{380}\ \overline{570})$, *all even numbers*, and as this plan worked for 35 and 55, why not for 95? So that must have been what he did! We cannot of course be sure that he did not search among the 3-term decompositions having even numbers; but if he had, he could have found, as KDF-9 shows us,

$$
\begin{array}{cccc}
A & \overline{56} & \overline{532} & \overline{760} \\
B & \overline{76} & \overline{152} & \overline{760} \\
C & \overline{76} & \overline{160} & \overline{608} \\
D & \overline{76} & \overline{190} & \overline{380} \\
E & \overline{80} & \overline{190} & \overline{304},
\end{array}
$$

---

* From his tables. See the EMLR or the G rule. Since $\overline{10}\ \overline{15} = \overline{6}, \overline{20}\ \overline{30} = \overline{12}$, and $\overline{380}\ \overline{570} = \overline{228}$, multiplying by 2 and 19.

and no doubt D and E would have tempted him, by virtue of Precept 1, while Precept 4 may have made him hesitate. We can scarcely say that the scribe made a bad choice for 2 ÷ 95; we can only say that he might have expressed his answer more concisely, and that it is a great pity he did not check with his tables.

2 ÷ 97   The only possible values for 97 are

A	$\overline{56}$	$\overline{679}$	$\overline{776}$	
B	$\overline{60}$	$\overline{679}$	$\overline{776}$	$\overline{840}$
C	$\overline{63}$	$\overline{504}$	$\overline{679}$	$\overline{776}$
D	$\overline{64}$	$\overline{448}$	$\overline{679}$	$\overline{776}$
E	$\overline{70}$	$\overline{280}$	$\overline{679}$	$\overline{776}$
F	$\overline{72}$	$\overline{252}$	$\overline{679}$	$\overline{776}$
G	$\overline{84}$	$\overline{168}$	$\overline{679}$	$\overline{776}$
H	$\overline{88}$	$\overline{154}$	$\overline{679}$	$\overline{776}.$

None is composed wholly of even numbers, and 776 is the least last number in all cases. Then A is clearly his best choice, for 56 is the smallest first term (Precept 4), and a 3-term decomposition is preferable to a 4-term one (Precept 2). The scribe therefore wrote $\overline{56}$ $\overline{679}$ $\overline{776}$ in the Recto Table.

2 ÷ 99   See 2 ÷ 9.

2 ÷ 101   There is only one possible decomposition for 2 divided by the prime number 101. It is $\overline{101}$ $\overline{202}$ $\overline{303}$ $\overline{606}$, which KDF-9 and the scribe both gave. It is derivable from the simple decomposition $\overline{2}$ $\overline{3}$ $\overline{6}$ = 1 that was very well known to the scribe. If this is rewritten as $\overline{1}$ $\overline{2}$ $\overline{3}$ $\overline{6}$ = 2, it is possible to produce a whole new Recto Table, consisting entirely of 4-term expressions,

$$
\begin{array}{llllll}
2 \div 3 = & \overline{3} & \overline{6} & \overline{9} & \overline{18} \\
2 \div 5 = & \overline{5} & \overline{10} & \overline{15} & \overline{30} \\
2 \div 7 = & \overline{7} & \overline{14} & \overline{21} & \overline{42} \\
2 \div 9 = & \overline{9} & \overline{18} & \overline{27} & \overline{54} \\
\end{array}
$$

. . . ,

of which the scribe was well aware, but which he did not want; too many terms, too many odd numbers!

The conclusion to which we are led regarding the Recto Table as a whole cannot be better expressed than in the words of P. Mansion (quoted on p. 48), *Summing up, all the decompositions of the Recto Table, from one point of view or another, are the very simplest of all the decompositions possible.*

E. M. Bruins* thinks that the only values given by the scribe which could have been improved upon are those for 2 divided by 13, 53, 61, 71, and 89, and he notes that $2 \div 95$ is reducible to a 2-term equality. He further notes that all except four of the 25 prime divisors follow the framework

$$2 \div p = \overline{N} \quad \overline{kp} \quad \overline{mp} \quad \overline{np},$$

where p is a prime and N, k, m, and n are integers. But in fact, the whole 25 are so expressible, with m and n both zero. Indeed, the whole fifty equalities follows this framework, where m and n may be zero, with slight variations for $2 \div 35$ and $2 \div 91$. I cannot agree that the scribal values for 2 divided by 13, 61, 71, and 89 could have been improved upon by the canon I have postulated.

* E. M. Bruins, "Ancient Egyptian Arithmetic," *Kon. Nederland Akademie van Wetenschappen*, Ser. A, Vol. 55, No. 2 (Amsterdam, 1952), pp. 81–91.

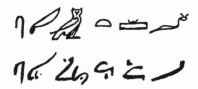

7   THE RECTO CONTINUED

EVEN NUMBERS IN THE RECTO: $2 \div 13$

The observation that the scribe preferred even numbers to odd numbers in his choice of equalities for the Recto of the RMP arose from pondering on the scribe's preference for $\overline{8}\ \overline{52}\ \overline{104}$ over $\overline{7}\ \overline{91}$ in the division of 2 by 13. For Precept 1 (page 49) calls for the smallest numbers, and Precept 2 says that a 2-term value is preferred to a 3-term value, so that at first glance one would have expected the scribe to have selected $\overline{7}\ \overline{91}$.

Clearly, the main purpose of the Recto Table is to facilitate divisions involving fractions. Thus the unit fraction values of 3, 4, 5, ..., 12 divided by 13 can be most expeditiously found from the even decompositions of $2 \div 13$, as I will now show. Thus we can find quite neatly the divisions

					from:
$1 \div 13$	$\overline{13}$				The Recto
$2 \div 13$	$\overline{8}$	$\overline{52}$	$\overline{104}$		
$3 \div 13$	$\overline{8}$	$\overline{13}$	$\overline{52}$	$\overline{104}$	$3 = 1 + 2$
$4 \div 13$	$\overline{4}$	$\overline{26}$	$\overline{52}$		$4 = 2 \times 2$
$5 \div 13$	$\overline{4}$	$\overline{13}$	$\overline{26}$	$\overline{52}$	$5 = 1 + 4$
$6 \div 13$	$\overline{4}$	$\overline{8}$	$\overline{13}$	$\overline{104}$	$6 = 2 + 4$
$7 \div 13$	$\overline{2}$	$\overline{26}$			$7 = 1 + 6$
$8 \div 13$	$\overline{2}$	$\overline{13}$	$\overline{26}$		$8 = 2 \times 4.$

We need not go past 8 at this stage, because from $9 \div 13$ onwards the fraction $\overline{\overline{3}}$ can be introduced, and so the technique changes slightly. Now we note that if the value $2 \div 13 = \overline{7}\ \overline{91}$ had been chosen, the previous table would have appeared as

					from:
$1 \div 13$	$\overline{13}$				
$2 \div 13$	$\overline{7}$	$\overline{91}$			
$3 \div 13$	$\overline{7}$	$\overline{13}$	$\overline{91}$		$3 = 1 + 2$
$4 \div 13$	$(\overline{4}$	$\overline{28})$	$(\overline{70}$	$\overline{130})$	$4 = 2 \times 2.$

At this stage the scribe would have had to stop, for although $2 \div 7 = \overline{4}\ \overline{28}$ had already been found, the equality $2 \div 91 = \overline{70}\ \overline{130}$ had not yet been determined. He would have been in the same predicament had he put $4 \div 13 = (1 + 3) \div 13$; in either case he would have been forced back to a tedious conventional division in order to evaluate $4 \div 13$:

	1		13	
	$\overline{2}$		6	$\overline{2}$
\	$\overline{4}$	\	3	$\overline{4}$
\	$\overline{26}$	\		$\overline{2}$
\	$\overline{52}$	\		$\overline{4}$

Totals $\overline{4}\ \overline{26}\ \overline{52}$ 4.

We conclude, therefore, that a preference for even numbers might well have been a very important factor in the preparation of the Recto Table as a whole. That there was a preference can be seen from an examination of Table 7.1.

We also observe this preponderance of even numbers in the following decompositions:

For   the scribe prefers

$2 \div 9$			$\overline{6}$	$\overline{18}$ to $\overline{5}$	$\overline{45}$	
$2 \div 33$			$\overline{22}$	$\overline{66}$,, $\overline{21}$	$\overline{77}$	
$2 \div 35$			$\overline{30}$	$\overline{42}$,, $\overline{21}$	$\overline{105}$	
$2 \div 55$			$\overline{30}$	$\overline{330}$,, $\overline{33}$	$\overline{165}$.	

$2 \div 13$		$\overline{8}$	$\overline{52}$	$\overline{104}$ to $\overline{7}$	$\overline{91}$	
$2 \div 41$		$\overline{24}$	$\overline{246}$	$\overline{328}$,, $\overline{21}$	$\overline{861}$	
$2 \div 71$		$\overline{40}$	$\overline{568}$	$\overline{710}$,, $\overline{42}$	$\overline{426}$	$\overline{497}$
$2 \div 95$		$\overline{60}$	$\overline{380}$	$\overline{570}$,, $\overline{57}$	$\overline{285}$.	

$2 \div 29$	$\overline{24}$	$\overline{58}$	$\overline{174}$	$\overline{232}$ to $\overline{29}$	$\overline{58}$	$\overline{87}$	$\overline{174}$
$2 \div 61$	$\overline{40}$	$\overline{244}$	$\overline{488}$	$\overline{610}$,, $\overline{61}$	$\overline{122}$	$\overline{183}$	$\overline{366}$
$2 \div 89$	$\overline{60}$	$\overline{356}$	$\overline{534}$	$\overline{890}$,, $\overline{89}$	$\overline{178}$	$\overline{267}$	$\overline{534}$
$2 \div 79$	$\overline{60}$	$\overline{237}$	$\overline{316}$	$\overline{790}$,, $\overline{79}$	$\overline{158}$	$\overline{237}$	$\overline{474}$.

In this group of decompositions preferred by the scribe, only *one* unit fraction is odd. In the complete table we find that the even unit

TABLE 7.1
Comparison of the frequency of appearance of even and odd unit fractions
in the Recto Table of the RMP.

	Number of equalities in the Recto Table	Number of 2-term equalities	Number of 3-term equalities	Number of 4-term equalities
	50	29	13	8
Number containing only even numbers	33	25	5	3
Number containing odd numbers	17	4	8	5
Total of numbers used	105 even, 24 odd, = 129	50 even, 8 odd, = 58	31 even, 8 odd, = 39	24 even, 8 odd, = 32

fractions exceed the odd by 5 to 1. We may therefore fairly include
this fifth precept in the scribes' canon for the RMP Recto Table.

In the light of the foregoing, consider the following statement by
F. Ll. Griffith:

Egyptian fractions were all primary, having numerators 1 except $2/3$.
Then to express $9/13$, they were obliged to reduce it thus.

$$9/13 = 2/13 + 2/13 + 2/13 + 2/13 + 1/13,$$
$$= 4 \times 2/13 + 1/13,$$
$$= 4(1/8 + 1/52 + 1/104) + 1/13, \qquad \text{(Recto, } 2 \div 13).$$
$$= 1/2 + 1/13 + 1/26 + 1/13,$$
$$= 1/2 + (1/13 + 1/13) + 1/26,$$
$$= 1/2 + (1/8 + 1/52 + 1/104) + 1/26, \quad \text{(Recto, } 2 \div 13).$$
$$9/13 = 1/2 + 1/8 + 1/26 + 1/52 + 1/104.*$$

Griffith continued, "This operation was performed in the head, only
the result being written down, and to facilitate it, tables were drawn
up of 2 divided by the odd numbers."† It is necessary to correct an
impression that might arise regarding the apparent complexity of
Egyptian fractions, if what Griffith offers as a decomposition of $9/13$

* From Griffith's article "Egypt," *Encyclopaedia Britannica*, 11th edition,
1910–1911.
† *Ibid.*

is in fact how the scribes would have done it. There can be no question as to Griffith's erudition as an Egyptologist, but his familiarity with classical Egyptian arithmetic must here be questioned. Any competent scribe would have found a much simpler value for $\frac{9}{13}$ in terms of unit fractions. Furthermore, he would have found it much more expeditiously than Griffith suggests, by the following simple division of 9 by 13:

$$
\begin{array}{ll}
\qquad 1 & \qquad 13 \\
\diagdown\ \bar{\bar{3}} & \diagdown\ 8 \quad \bar{\bar{3}} \\
\qquad 3 & \qquad 39 \\
\diagdown\ \overline{39} & \diagdown \qquad \bar{3} \\
\hline
\text{Totals } \bar{\bar{3}} \quad \overline{39} & \qquad 9.
\end{array}
$$

Then, $\frac{9}{13} = \bar{\bar{3}}\ \overline{39}$. Or again, the scribe may have divided a little differently:

$$
\begin{array}{ll}
\qquad 1 & \qquad 13 \\
\diagdown\ \bar{2} & \diagdown\ 6 \quad \bar{2} \\
\qquad \bar{3} & \qquad 4 \quad \bar{3} \\
\diagdown\ \bar{6} & \diagdown\ 2 \quad \bar{6} \\
\diagdown\ \overline{39} & \diagdown \qquad \bar{3} \\
\hline
\text{Totals } \bar{2} \quad \bar{6} \quad \overline{39} & \qquad 9.
\end{array}
$$

Then, $\frac{9}{13} = \bar{2}\ \bar{6}\ \overline{39}$. These two simple reductions to two-term and three-term equalities serve to show that, despite the restriction to unit numerators imposed by their notation, the Egyptians developed a powerful technique for such reductions, not wholly dependent upon the table of the RMP Recto, as Griffith implies. Furthermore, they have a greater range of answers according to their choice of multipliers, so that they could obtain the 4-term equality $\frac{9}{13} = \bar{2}\ \bar{8}\ \overline{16}\ \overline{208}$, and even a 5-term decomposition $\frac{9}{13} = \bar{2}\ \overline{13}\ \overline{14}\ \overline{35}\ \overline{65}$, but with smaller denominators than those found by Griffith. With the Egyptian's well-known preference for the fraction $\bar{\bar{3}}$ whenever possible, I suggest the scribe's value for $\frac{9}{13}$ would have been $\bar{\bar{3}}\ \overline{39}$.

MULTIPLES OF DIVISORS IN THE RECTO

The first entry in the Recto is $2 \div 3 = \bar{\bar{3}}$, which could have been written, as the scribe well knew, as

$$2 \div 3 = \bar{2} \quad \bar{6}.$$

Now there are 17 divisions of 2 by the divisors which are multiples of 3, namely, 3, 9, 15, . . . 99; these are all obtainable from the first equality on multiplying through by 3, 5, 7, . . . , 33, giving the entries that are in the RMP:

2 divided by	resulting equality		2 divided by	resulting equality	
3	$\bar{2}$	$\bar{6}$	57	$\overline{38}$	$\overline{114}$
9	$\bar{6}$	$\overline{18}$	63	$\overline{42}$	$\overline{126}$
15	$\overline{10}$	$\overline{30}$	69	$\overline{46}$	$\overline{138}$
21	$\overline{14}$	$\overline{42}$	75	$\overline{50}$	$\overline{150}$
27	$\overline{18}$	$\overline{54}$	81	$\overline{54}$	$\overline{162}$
33	$\overline{22}$	$\overline{66}$	87	$\overline{58}$	$\overline{174}$
39	$\overline{26}$	$\overline{78}$	93	$\overline{62}$	$\overline{186}$
45	$\overline{30}$	$\overline{90}$	99	$\overline{66}$	$\overline{198}.$
51	$\overline{34}$	$\overline{102}$			

After one or two of these multiplications, it would become evident that these three columns of numbers form three series, so that it would be easier to keep adding 6 to the left-hand sides and then 4 and 12 to the two right-hand columns. Having his answers already prepared, they are proved to be correct in exactly the same way, as illustrated, for example, by 2 ÷ 51 from the Recto:

$$
\begin{array}{cc}
1 & 51 \\
\bar{3} & 34 \\
\diagdown\ \overline{34} & \diagdown\ 1 \quad \bar{2} \\
\diagdown\overline{102} & \diagdown \quad\quad \bar{2} \\
\hline
\text{Totals } \overline{34} \quad \overline{102} & \quad 2.
\end{array}
$$

This same sequence for this particular group of 17 divisions is also obtainable using the rule given in RMP 61B (page 29), and it is quite possible that the scribe obtained his answers this way. Thus

$$2 \div\ \ 3 = \bar{\bar{3}} \text{ of } 1 = \overline{2 \times 1} + \overline{6 \times 1} = \ \bar{2} \quad\ \bar{6}$$
$$2 \div\ \ 9 = \bar{\bar{3}} \text{ of } 3 = \overline{2 \times 3} + \overline{6 \times 3} = \ \bar{6} \quad\ \overline{18}$$
$$2 \div 15 = \bar{\bar{3}} \text{ of } 5 = \overline{2 \times 5} + \overline{6 \times 5} = \overline{10} \quad \overline{30}$$
$$2 \div 21 = \bar{\bar{3}} \text{ of } 7 = \overline{2 \times 7} + \overline{6 \times 7} = \overline{14} \quad \overline{42}$$

. . . .

There are 7 divisions of 2 by multiples of 5 (omitting those like 15, which is also a multiple of 3), and from $2 \div 5 = \overline{3}\ \overline{15}$ we obtain similarly by continuous multiplication the equalities

$2 \div 5 =$	$\overline{3}$	$\overline{15}$		
$2 \div 25 =$	$\overline{15}$	$\overline{75}$	$(\times\ 5)$	
$2 \div 35 =$	$\overline{21}$	$\overline{105}$	$(\times\ 7)$	(but Recto has $\overline{30}\ \overline{42}$)
$2 \div 55 =$	$\overline{33}$	$\overline{165}$	$(\times\ 11)$	(but Recto has $\overline{30}\ \overline{330}$)
$2 \div 65 =$	$\overline{39}$	$\overline{195}$	$(\times\ 13)$	
$2 \div 85 =$	$\overline{51}$	$\overline{255}$	$(\times\ 17)$	
$2 \div 95 =$	$\overline{57}$	$\overline{285}$	$(\times\ 19)$	(but Recto has $\overline{60}\ \overline{380}\ \overline{570}$).

We have already noted Aᶜh-mosè's poor choice in $2 \div 95$. In a similar manner we find for multiples of 7,

$2 \div 7 =$	$\overline{4}$	$\overline{28}$		
$2 \div 49 =$	$\overline{28}$	$\overline{196}$	$(\times\ 7)$	
$2 \div 77 =$	$\overline{44}$	$\overline{308}$	$(\times\ 11)$	
$2 \div 91 =$	$\overline{52}$	$\overline{364}$	$(\times\ 13)$	(but Recto has $\overline{70}\ \overline{130}$).

The multiples of 11, namely 33, 55, 77, and 99, have already been considered as multiples of 3, 5, and 7; of the multiples of 13, namely 39, 65, and 91, there remains only 91 so far unconsidered. Then,

$$2 \div 13 = \overline{8}\quad \overline{52}\quad \overline{104}, \qquad \text{or } \overline{7}\quad \overline{91}$$
$$2 \div 91 = \overline{56}\quad \overline{364}\quad \overline{728}\quad (\times\ 7), \text{ or } \overline{49}\quad \overline{637}\quad (\times\ 7).$$

But the Recto has $2 \div 91 = \overline{70}\ \overline{130}$.

We need go no further, because higher multiples of 17, 19, 23, 29, etc., have already been considered or are greater than 101. It will be noticed that for $2 \div 91$ the scribe entered a decomposition superior to either one found by multiplication of earlier expressions of $2 \div 13$. There is a further method by which the equalities for divisors which are multiples of 3 could have been found, that utilizes the tables of 2-term unit fractions:

$$2 \div 3 = \overline{3} \qquad \overline{3}$$
$$= \overline{3}\quad (\overline{6}\quad \overline{6})$$
$$= (\overline{3}\quad \overline{6})\quad \overline{6}$$
$$= \quad \overline{2}\quad \overline{6};$$

$$
\begin{aligned}
2 \div 9 &= \bar{9} & & \bar{9} \\
&= \bar{9} & (\overline{18} & \quad \overline{18}) \\
&= (\bar{9} & \overline{18}) & \quad \overline{18} \\
&= \bar{6} & & \quad \overline{18};
\end{aligned}
$$

$$
\begin{aligned}
2 \div 15 &= \overline{15} & & \overline{15} \\
&= \overline{15} & (\overline{30} & \quad \overline{30}) \\
&= (\overline{15} & \overline{30}) & \quad \overline{30} \\
&= \overline{10} & & \quad \overline{30};
\end{aligned}
$$

$$
\begin{aligned}
2 \div 57 &= \overline{57} & & \overline{57} \\
&= \overline{57} & (\overline{114} & \quad \overline{114}) \\
&= (\overline{57} & \overline{114}) & \quad \overline{114} \\
&= \overline{38} & & \quad \overline{114}.
\end{aligned}
$$

This technique can be repeated up to $2 \div 99$.

TWO DIVIDED BY THIRTY-FIVE:
THE SCRIBE DISCLOSES HIS METHOD

In every one of the fifty divisions of the Recto, the scribe wrote his equality on the top line in contrasting colors, the fractions constituting the answers being in red, and all the rest in black. *In 2 ÷ 35 alone the scribe added one further explanatory line*, again in contrasting red and black, which gives us the clue to his method, at least for the divisor 35. (What we have here is, in fact, an example of the scribe's use of the so-called red auxiliaries. In order to understand the division 2 ÷ 35 we must anticipate the discussion of Chapter 8, where the uses of the red auxiliaries are treated in detail.)

The first line of Figure 7.1 shows the two fractions 30 and 42 in red (reading from right to left). In the second line, there is first the number 6 in red, underneath the black divisor 35; this is followed by the numbers 7 and 5 in black, placed underneath the red fractions $\overline{30}$ and $\overline{42}$, respectively. It is this second line, *which never occurs in any other division of the Recto*, that enables us to understand and explain how the scribe arrived at the answer of $\overline{30}\ \overline{42}$ for the division of 2 by 35. We rewrite 2 ÷ 35 in modern form from left to right, using boldface type for those numbers that the scribe drew in red:

$$35 \quad \overline{30} \text{ of } 35 = 1 \quad \overline{6} \quad \overline{42} \text{ of } 35 = \overline{\overline{3}} \quad \overline{6}$$
$$\mathbf{6} \qquad 7 \qquad\qquad\qquad 5.$$

By this the scribe means that since

$$(\overline{30} \text{ of } 35) + (\overline{42} \text{ of } 35) = (1 + \overline{6}) + (\overline{\overline{3}} + \overline{6})$$
$$= 2,$$

then

$$\overline{30} + \overline{42} = 2 \div 35.$$

The red 6 placed underneath the divisor 35 is the number he has chosen as the most convenient multiplier of 35, giving 210 which we may regard as the Egyptian counterpart of our modern "common multiple," which may or may not be at the same time the "least common multiple" so familiar to us. Then somewhere or other, perhaps on a sort of memorandum papyrus pad, the scribe multiplied by 2 this red 6, giving 12, which, in turn, he had to partition into two, three, or perhaps more parts that will each divide 210 *without remainder*. In this

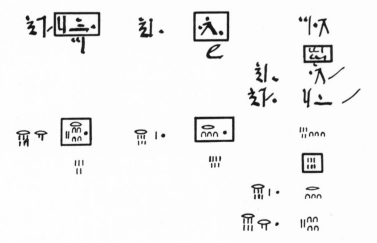

FIGURE 7.1
The division 2 ÷ 35 from the RMP. The red auxiliaries of the original are shown here enclosed in boxes. The hieroglyphic transliteration is shown below the hieratic version. Read from right to left.

case, he found that he could do this with two parts, i.e., he found $12 = 7 + 5$, with $210 \div 7 = 30$ and $210 \div 5 = 42$, and these are the numbers, expressed as unit fractions, that he wrote in red above the 7 and 5 in the papyrus. Then in his two lines of proof he shows that $\overline{30}$ of 35 is indeed $1\ \overline{6}$, and that $\overline{42}$ of 35 is $\bar{\bar{3}}\ \overline{6}$. Of course the number 12 could have been partitioned differently, as, for example, $12 = 10 + 2$, and the resulting equality would have been $\overline{21} + \overline{105} = 2 \div 35$. No doubt the scribe tried this and rejected it, preferring the former equality. No other partitioning of 12 will give two numbers which will divide 210 without remainder, but there could be other partitionings into three or four parts, such as $12 = 7 + 3 + 2$, leading to $\overline{30} + \overline{70} + \overline{105} = 2 \div 35$, which the scribe would certainly have rejected, had it come to his attention.

It is tempting to conclude that the scribe's technique as shown in RMP $2 \div 35$ was his standard method in all fifty divisions of the Recto. But this was not the case, as a close examination of the various divisions discloses; it appears that wherever possible he used it, yet as he progressed he treated each division on its merits, finding real difficulties with the prime divisors 29, 43, 61, 73, 83, and 89. The somewhat controversial division of 2 by 13 illustrates one of his different attacks in the divisions of the Recto. This is the calculation of $2 \div 13$ as it occurs in the Recto Table.

$$13 \quad \mathbf{\overline{8}}\ \text{of } 13 = 1\ \overline{2}\ \overline{8} \quad \mathbf{\overline{52}}\ \text{of } 13 = \overline{4} \quad \mathbf{\overline{104}}\ \text{of } 13 = \overline{8}$$

```
                   1                    13
                   2̄                     6    2̄
                   4̄                     3    4̄
                  \8̄                    \1    2̄    8̄
                 \ 52̄                   \           4̄
                 \104̄                   \                 8̄
         ───────────────────          ─────────────────────────
   Totals 8̄     52̄     104̄              1    2̄    4̄(8̄   8̄)
                                           = 2.
```

Had the scribe alternatively commenced with $\bar{\bar{3}}$, then $\bar{\bar{3}}$, $\overline{6}$, $\overline{12}$, ..., he would have found the more cumbersome equality

$$\overline{12} \quad \overline{26} \quad \overline{52} \quad \overline{78} = 2 \div 13,$$

which the reader may care to deduce in detail for himself. No doubt he tried both, and preferred the first equality, while the equality,

$$\overline{7} \quad \overline{91} = 2 \div 13,$$

did not come to his attention; we can state this with some certainty, because he could not take $\overline{7}$ of 13 unless he did the separate division of $6 \div 7$, and this results in a quite extensive calculation, which he would not attempt.

8 PROBLEMS IN COMPLETION AND THE RED AUXILIARIES

USE OF THE RED AUXILIARIES OR REFERENCE NUMBERS

Problems 21, 22, and 23 of the RMP are called *problems in completion*, since the scribe writes "complete $\bar{3}$ $\overline{15}$ to 1," when he means "subtract $\bar{\bar{3}}$ $\overline{15}$ from 1," which could also be expressed as, "What must be added to $\bar{\bar{3}}$ $\overline{15}$ to make 1?" In these three problems, he shows how to use a *reference number*,* which is the Egyptian counterpart of our least common denominator, for handling various fractions. But the Egyptians did not worry about the reference number being the least which could be chosen. They usually chose the highest number of the fractions before them, but not always. And in the calculations that followed, they wrote what would correspond to our numerators in red instead of black. For this reason, these numbers are sometimes referred to as *red auxiliaries*.

The use of the red auxiliaries was so common in Egyptian computation that no scribe's palette was made without at least one extra depression for red pigment (Figure 8.1, bottom). Indeed, the two-depression palette came to be regarded as a symbol of the scribe's office, and the hieroglyph for "scribe" was a stylized palette and writing stick (Figures 8.1, top and 8.2).

In the workings that follow, the red auxiliaries are set in boldface type. I have added solutions to these three problems, using unit-fraction equalities from tables that the scribe might have used if he had not had another specific pedagogic purpose in view.

PROBLEM 21

Complete $\bar{\bar{3}}$ $\quad$ $\overline{15}$ to 1 $\quad$ [Answer: $\bar{5}$ $\quad$ $\overline{15}$].

Take 15 as a reference number, and use the red auxiliaries: $\bar{\bar{3}}$ of 15 is 10, $\overline{15}$ of 15 is 1, and 1 part of 15 is 15. Then,

$$\bar{\bar{3}} \quad \overline{15} + \text{(some other fractions)} = 1$$
$$\mathbf{10} \quad \mathbf{1} \quad (\qquad \mathbf{4} \qquad) = \mathbf{15}.$$

* Called *Hilfszahlen* by Vogel. See his Vorgriechische Mathematik, Vol. 1, *Vorgeschichte und Ägypten*, Schroedel, Hanover, 1958, p. 40.

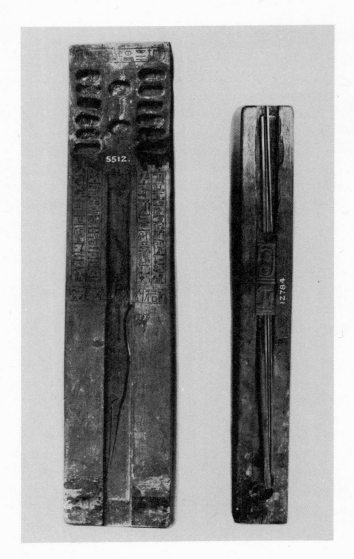

FIGURE 8.1
Two scribes' wooden palettes. *Left*, palette inscribed with the name of King Tuthmosis IV (1425–1417 B.C.) of the Eighteenth Dynasty. The inscriptions along the sides are funerary invocations for the high official Meryre, followed by the scribe's name (Tunen). Length 13 inches. *Right*, two-depression palette with a number of writing instruments. The inscription is the name of King Amosis (1570–1546) of the Eighteenth Dynasty. Length 11¼ inches. Courtesy British Museum.

FIGURE 8.2
A wooden panel from the tomb of Hesy-Ra, showing a scribe with the insignia of his office over his shoulder and writing materials in his hand. Courtesy Cairo Museum.

We have to find what fraction or fractions of 15 will give 4. Divide 4 by 15.

$$
\begin{array}{ccc}
1 & & 15 \\
\overline{10} & & 1 \quad \overline{2} \\
\diagdown \; \overline{5} & & \diagdown \; 3 \\
\diagdown \overline{15} & & \diagdown \; 1 \\
\hline
\text{Totals} \; \overline{5} \quad \overline{15} & & 4.
\end{array}
$$

Therefore $\overline{5}\ \overline{15}$ is needed to complete $\overline{\overline{3}}\ \overline{15}$ to 1. (Note the doubling of $\overline{10}$ to obtain $\overline{5}$.)

PROOF

$$
\begin{array}{cccc}
\overline{\overline{3}} & \overline{5} & \overline{15} & \overline{15} \text{ makes } 1 \\
\mathbf{10} & \mathbf{3} & \mathbf{1} & \mathbf{1} \text{ makes } \mathbf{15.}
\end{array}
$$

Alternatively, the scribe might have said,

$$
\begin{array}{lll}
1 = \overline{\overline{3}} & & \overline{\overline{3}} \\
\; = \overline{\overline{3}} & \overline{6} & \overline{6} & \text{gen.* } (1, 1) \\
\; = \overline{\overline{3}} \; (\overline{10} \quad \overline{15}) & & \overline{6}, & \text{gen. } (2, 3) \\
\; = \overline{\overline{3}} \quad \overline{15} & (\overline{6} \quad \overline{10})
\end{array}
$$

therefore,

$$
1 = \overline{\overline{3}} \quad \overline{15} \; (\overline{6} \; \overline{10}),
$$

or $\overline{6}\ \overline{10}$ is needed to complete $\overline{\overline{3}}\ \overline{15}$ to 1.

PROBLEM 22

Complete $\overline{\overline{3}}\quad \overline{30}$ to 1   [Answer: $\overline{5}\quad \overline{10}$].

Take 30 as a reference number, and use the red auxiliaries: $\overline{\overline{3}}$ of 30 is 20, $\overline{30}$ of 30 is 1, and 1 part of 30 is 30. Then,

$$
\begin{array}{lll}
\overline{\overline{3}} \quad \overline{30} + \text{(some other fractions)} & = & 1 \\
\mathbf{20} \quad \mathbf{1} \; (\quad\quad \mathbf{9} \quad\quad) & = & \mathbf{30.}
\end{array}
$$

* See p. 104 for the definition and use of the generator.

We have to find what fraction or fractions of 30 will give 9. Divide 9
by 30.

$$
\begin{array}{cc}
1 & 30 \\
\diagdown\,\overline{10} & \diagdown\ 3 \\
\diagdown\ \overline{5} & \diagdown\ 6 \\
\hline
\text{Totals }\ \overline{5}\quad \overline{10} & 9.
\end{array}
$$

Therefore $\overline{5}\ \overline{10}$ is needed to complete $\overline{3}\ \overline{30}$ to 1. (Again $\overline{10}$ is doubled
to obtain $\overline{5}$.)

PROOF

$$
\begin{array}{ccccl}
\overline{\overline{3}} & \overline{5} & \overline{10} & \overline{30} & \text{makes}\ \ 1 \\
\mathbf{20} & \mathbf{6} & \mathbf{3} & \mathbf{1} & \text{makes}\ \mathbf{30.}
\end{array}
$$

Alternatively, the scribe might have noted from Problem 21 that
$\overline{3}\ \overline{30}$ needs $\overline{30}$ more than $\overline{3}\ \overline{15}$ to make 1. Then $\overline{6}\ \overline{10}\ \overline{30}$ must be
needed. But

$$
\begin{aligned}
\overline{6}\quad \overline{10}\quad \overline{30} &= (\overline{6}\quad \overline{30})\quad \overline{10} \\
&= \quad\ \ \overline{5}\qquad\quad \overline{10}. \qquad\qquad \text{gen. } (1, 5)
\end{aligned}
$$

Or, he may have noted from the Recto, $2 \div 91, \overline{\overline{3}}\ \overline{5}\ \overline{10}\ \overline{30} = 1$.

PROBLEM 23

  Complete $\overline{4}\quad \overline{8}\quad \overline{10}\quad \overline{30}\quad \overline{45}$ to $\overline{\overline{3}}$ [Answer: $\overline{9}\quad \overline{30}$].

Take 45 as a reference number and use red auxiliaries. Then,

$$
\begin{array}{ccccc}
\overline{4} & \overline{8} & \overline{10} & \overline{30} & \overline{45}\ + \text{(some other fractions)} = \overline{\overline{3}} \\
(\mathbf{11}\ \overline{4} & \mathbf{5}\ \overline{2}\ \overline{8} & \mathbf{4}\ \overline{2} & \mathbf{1}\ \overline{2} & \mathbf{1})\ + \text{(some other fractions)} = \mathbf{30} \\
& \mathbf{23}\ \overline{2}\ \overline{4}\ \overline{8} & & \mathbf{6}\ \overline{8} & = \mathbf{30.}
\end{array}
$$

We have to find what fractions or fractions of 45 will give $6\ \overline{8}$. Divide
$6\ \overline{8}$ by 45.*

* The scribe does not show this division in Problem 23. In Problem 21 the
scribe forgot to change to red ink for the auxiliaries, and in Problem 22
he used no red ink at all. This is most unusual. RMP 79 is the only other
instance where the statement of the problem is not in red.

1	45	
$\overline{3}$	15	
\ $\overline{9}$	\ 5	
$\overline{10}$	4 $\overline{2}$	
$\overline{20}$	2	$\overline{4}$
\ $\overline{40}$	\ 1	$\overline{8}$
Totals $\overline{9}$ $\overline{40}$	6 $\overline{8}$.	

Therefore $\overline{9}$ $\overline{40}$ is needed to complete $\overline{4}$ $\overline{8}$ $\overline{10}$ $\overline{30}$ $\overline{45}$ to $\overline{3}$.

PROOF

$\overline{4}$	$\overline{8}$	$\overline{9}$	$\overline{10}$	$\overline{30}$	$\overline{40}$	$\overline{45}$ and $\overline{3}$ make 1
11 $\overline{4}$	**5 $\overline{2}$ $\overline{8}$**	**5**	**4 $\overline{2}$**	**1 $\overline{2}$**	**1 $\overline{8}$**	**1** and **15** make **45**.

Alternatively,

$$\overline{\overline{3}} = \quad \overline{3} \qquad \overline{6} \qquad \overline{6},$$
$$= (\overline{4}\ \overline{12})\ (\overline{8}\ \overline{24})\ (\overline{9}\ \overline{18}), \qquad \text{gen. (1, 3) and (1, 2)}$$
$$= (\overline{4}\ \overline{8}\ \overline{9})\ (\overline{12}\ \overline{24})\ \overline{18},$$
$$= (\overline{4}\ \overline{8}\ \overline{9})\ \overline{8}\ (\overline{30}\ \overline{45}), \qquad \text{gen. (1, 2) and (2, 3)}$$
$$= (\overline{4}\ \overline{8}\ \overline{9})\ (\overline{10}\ \overline{40})\ (\overline{30}\ \overline{45}), \qquad \text{gen. (1, 4)}$$
$$\overline{\overline{3}} = (\overline{4}\ \overline{8}\ \overline{10}\ \overline{30}\ \overline{45})\ (\overline{9}\ \overline{40}),$$

so $(\overline{9}\ \overline{40})$ is needed.

AN INTERESTING OSTRACON

The generally accepted meaning of the word "ostracon" is an inscribed fragment of pottery with Egyptian, Coptic, or Greek inscriptions. Ostraca are most commonly found in Upper Egypt, and date roughly from 600 B.C. to A.D. 400. The name comes from the Greek οστρακον, potsherd. Broken pieces of pottery had many uses in the ancient world, one of which was as a substitute for expensive papyrus if only brief notes needed to be made and if the clay had not previously been decorated or inscribed.

In the tomb of Sen-mut, an architect for Queen Hatshepsut (1520–1480 B.C.), several ostraca were found. Sen-mut designed the temple at Deir el-Bahri, thought by many to be the finest in ancient Egypt,*

* E. K. Milliken, *Cradles of Western Civilisation*, Harrap, London, 1955, p. 92.

for the Queen. Sen-mut's tomb is at Thebes, where the New York
Metropolitan Museum of Art conducted an expedition, obtaining an
arithmetic computation on an ostracon that was subsequently translated.* This computation consists of three double lines in red and
black, showing the answer to the divisions of 2 and 4 by 7, expressed
in unit fractions. Now 2 divided by 7 is one of the fifty divisions of the
RMP Recto, where the standard answer is recorded as

$$2 \div 7 = \overline{4} \quad \overline{28}.$$

The interesting thing about this ostracon is that the decomposition
calculated on it is the unexpected 3-term one

$$2 \div 7 = \overline{6} \quad \overline{14} \quad \overline{21},$$

and the author's technique in deriving it throws some light upon the
Egyptian method of using a reference number together with red
auxiliaries when adding fractions. The computation, with boldface
type indicating the red numbers, is

l. 1	1	$\overline{7}$		
l. 2		**3**		
l. 3	2	$\overline{6}$	$\overline{14}$	$\overline{21}$
l. 4		**3 $\overline{2}$**	**1 $\overline{2}$**	**1**
l. 5	4	$\overline{2}$	$\overline{14}$	
l. 6		**10 $\overline{2}$**	**1 $\overline{2}$.**	

Whoever inscribed the ostracon was doing just what the scribe of the
RMP did in problems 28, 32, 36 and several others. The red 3 beneath
the $\overline{7}$ means "Take 3 as a multiplier of 7, to give the reference number
21." He then multiplied the 2 (of line 3) by his multiplier to give 6
which he then partitioned as $3\frac{1}{2}$, $1\frac{1}{2}$, and 1, each of which divides
the reference number 21 in integers, and wrote them in red (line 4).
These are the red auxiliaries.

The scribe of the ostracon then referred these auxiliaries to 21,
finding that $3\frac{1}{2}$ is $\frac{1}{6}$ of 21, $1\frac{1}{2}$ is $\frac{1}{14}$ of 21, and 1 is $\frac{1}{21}$ of 21, so that

---

* See W. C. Hayes, "Ostracon No. 153," *Ostraca and Name Stones from the
Tomb of Sen-mut at Thebes*, Publication 15, Metropolitan Museum of Art,
New York, 1942.

he wrote $\bar{6}$, $\overline{14}$, $\overline{21}$ in black in their proper places (line 3). In this terse manner the scribe obtained his answer to the division,

$$2 \div 7 = \bar{6} \quad \overline{14} \quad \overline{21}.$$

Still with the same red multiplier 3 and the same reference number 21, we note the 4 (of line 5) was multiplied by 3 giving 12 which was partitioned as $10\frac{1}{2}$ and $1\frac{1}{2}$, each of which divides the reference number 21 in integers 2 and 14, which he wrote as $\bar{2}$ and $\overline{14}$ in their proper places (line 5) in black. It was thus he obtained his answer to the division,

$$4 \div 7 = \bar{2} \quad \overline{14}.$$

We have no way of telling how the scribe came to choose 3 as his multiplier* and, consequently, 21 as his reference number. Nor do we know how he decided upon his particular partitions of 6 and 12. This is part of his art, learned no doubt through past experience and constant practice. The method of the ostracon is a powerful one and, needing to write down only one line for each, the scribe could have extended the calculation to obtain,

$$3 \div 7 = \bar{4} \quad \bar{7} \quad \overline{28}$$
$$5 \div 7 = \bar{2} \quad \bar{7} \quad \overline{14}$$
$$6 \div 7 = \bar{3} \quad \bar{7} \quad \overline{21},$$

which the reader is invited to verify as the scribe would have done it.

* See O. Neugebauer, *The Exact Sciences in Antiquity*, Harper Torchbooks, Harper, New York, 1962, pp. 92–94, for another discussion of Ostracon 153.

9 THE EGYPTIAN MATHEMATICAL LEATHER ROLL

The Egyptian Mathematical Leather Roll (EMLR) was purchased together with the RMP by the Scotsman A. H. Rhind (1833–1863) at Luxor, Egypt, in 1858. This young lawyer came to Egypt's mild climate for reasons of health; he became interested in Egyptology, and went to Thebes in 1855, where he specialized in tombs. The EMLR and the RMP were discovered in some ruins of the Rameseum at Thebes, and later were acquired by Rhind. The papyri came to the British Museum in 1864, and have remained there ever since. The EMLR is roughly 10 inches by 17 inches; because of its very brittle condition, it remained unrolled for more than 60 years.

Dr. Alexander Scott and H. R. Hall finally succeeded in unrolling it in 1927,* and it was found to contain a collection, in duplicate, of 26 sums done in unit fractions. The lucky circumstance of duplication helped to make possible the restoration of the right-hand column, parts of which had been damaged. Various opinions as to its significance have been advanced. S. R. K. Glanville† regarded it as a "handy table for popular use," probably, he thought, the work of a junior official, "not of a schoolboy for the writing is far too good." That it might have been an answer to an examination paper seemed to him not to be the case, because of the duplicate copy. However, he felt sure it must have been copied from a textbook as a practical guide or table for future work. He goes on,

Its real mathematical interest lies in discovering what would have been the use of such a table to a person armed with it, and further, what was its relation, if any, to the Rhind Papyrus with which it was discovered? Such an inquiry must follow a more detailed discussion of the text itself.

From the scientific point of view, it can hardly be denied that the dissemination of the knowledge of the chemical treatment of the

* A. Scott and H. R. Hall, "Egyptian Leather Roll of the 17th Century B.C.," *British Museum Quarterly*, Vol. 2 (1927), pp. 56f.
† S. R. K. Glanville, "The Mathematical Leather Roll in the British Museum," *Journal of Egyptian Archaeology*, Vol. 23 (1927), pp. 237–239.

leather', is of greater value than the publication of the contents inscribed on it.*

In like vein, Scott and Hall express the opinion that

The roll has not justified the hope that it might prove to contain material of importance. It is simply a series of sums in additions of fractions, repeated twice over, apparently a scholar's exercise.†

A much more optimistic opinion as to the value of the contents of the EMLR was held by Vogel,‡ who wrote

Ich möchte den Inhalt als höchst bedeutsam ansehen, trotzdem er lediglich in 26 Stammbruchsummen besteht. (I consider the content as most important, although it consists only of 26 unit fractions.)

And Neugebauer similarly regards the EMLR as of much greater significance; he wrote§

Ich möchte, im folgenden zu zeigen versuchen, dass ein so pessimistisches Urteil vielleicht doch nicht ganz am Platze ist. (I will try to show in what follows, that such a pessimistic judgment would perhaps not be appropriate.)

When after sixty years the contents of the EMLR became known, the disappointment of Glanville and Scott and Hall is quite understandable. They, naturally enough, expected that any hieratic writing important enough to warrant inscription on costly leather, instead of the much more common and cheaper papyrus, would excite Egyptologists far beyond what previous historical disclosures had done. It might have spoken, for example, of the construction of the pyramids, of the temple of Karnak, of Abu Simbel; it might have contained the words of some famous pharaoh, perhaps Ramses, Akhenaton, or

* *Ibid.*
† Scott and Hall, "Egyptian Leather Roll."
‡ K. Vogel, "Erweitert die Lederolle unserer Kenntniss ägyptischer Mathematik?" *Archiv für Geschichte der Mathematik*, Vol. 2 (1929), pp. 386–407.
§ O. Neugebauer, "Zur ägyptischen Bruchrechnung," *Zeitschrift für Ägyptische Sprache*, Leipzig, Vol. 64 (1929), pp. 44f. See also B. L. Van der Waerden, "Die Entstehungsgeschichte der ägyptischen Bruchrechnung," *Quellen und Studien zur Geschichte der Mathematik*, Part B, Study IV, Berlin, 1937–1938, pp. 359–382.

Cheops; or recorded something of the diplomatic situation with the Babylonians or Akkadians, comparable perhaps with the Tell el ᶜAmarna tablets; indeed it might have shed light on any one of the many questions with which Egyptologists had been occupied for over a century. But it did none of these things.

A reconstruction and translation of the EMLR is given in Figure 9.1; a photograph of the roll itself is given in Figure 9.2.

However the table came to be written, let us imagine with Glanville that it was the work of a junior official, and that a chief scribe, perhaps the teacher, was required to show it to the head of the school for his approval. A kind of supervisor's inspection of work, so to speak. We can imagine the examiner's comments.

. . . His writing has improved, though the tails on his hundreds are rather long. And he has not arranged the equalities at all systematically. Observe some very simple ones mixed up with much harder ones. Look at number 8 for instance! I'm glad to see number 12 there, one of the most useful of all. And I detect a major error in line 17. Surely that should read $\overline{26}\ \overline{39}\ \overline{78} = \overline{13}$. It stares one in the face, from $\overline{2}\ \overline{3}\ \overline{6} = 1$. However, only one mistake is not bad. Of course the last eight lines are pretty repetitious, and should have included numbers 11 and 13 among them. On the whole, a good piece of work, and a very useful table for the addition of fractions. . . .

These comments are of course the comments we ourselves would make, but more importantly from the modern point of view the EMLR table throws great light upon the mechanical arithmetic of the RMP, the MMP,* the KP,† RP,‡ BP,§ AMP,‖ MichP.,# and many of the later Greek and Byzantine tables of fractions, as well as offering justification

* Moscow Mathematical Papyrus (originally, Golenischev Papyrus). Moscow Museum of Fine Arts. Inventory No. 4576.
† Kahun Papyrus. British Museum. Found by W. M. F. Petrie (grandson of the famous Australian explorer, Matthew Flinders) at Kahun, Egypt, in 1889. The papyrus contains six mathematical fragments, not all of which have been penetrated. See p. 176.
‡ Reisner Papyri. Museum of Fine Arts, Boston. Acquisition No. 38.2062.
§ Berlin Papyrus. Staatliche Museen, Berlin. No. 6619.
‖ Akhmîm Papyrus. Cairo Museum. No. 10,758 in the Catalogue of Papyri.
Michigan Papyri. University of Michigan, Ann Arbor. Vol. III of the University of Michigan Collection.

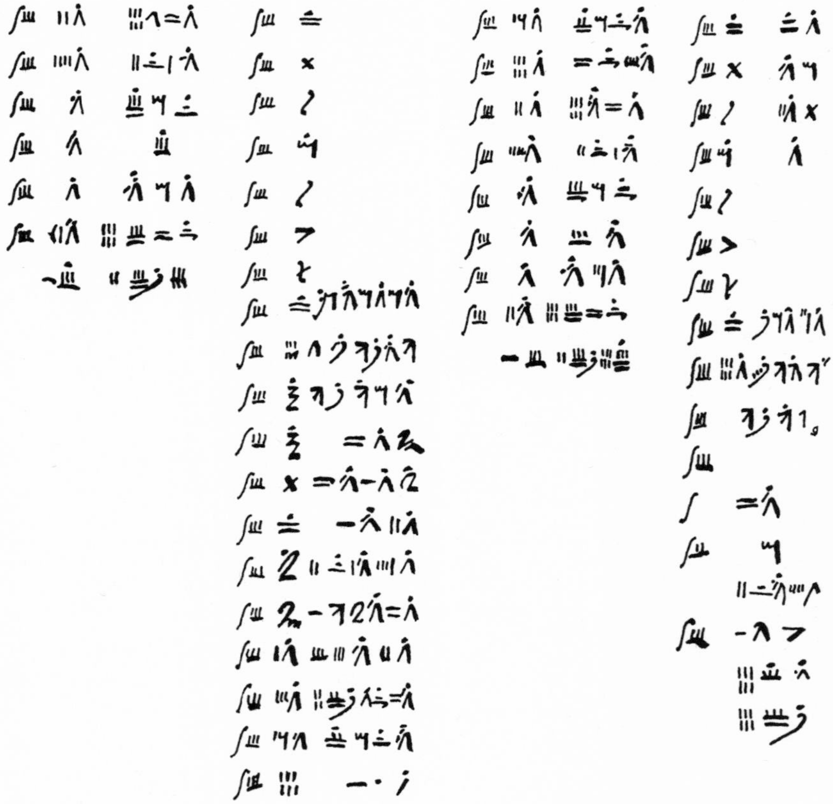

FIGURE 9.1
Left, reconstruction of the EMLR; *right*, Glanville's translation. Courtesy
S. R. K. Glanville.

1	10	40			8
2	5	20			4
3	4	12			3
4	10	10			5
5	6	6			3
6	6	6	6		2
7	3	3			$\frac{2}{3}$
8	25	15	75	200	8
9	50	30	150	400	16
10	25	50	150		15
11	9	18			6
12	7	14	28		4
13	12	24			8
14	14	21	42		7
15	18	27	54		9
16	22	33	66		11
17	28	49	196		13
18	30	45	90		15
19	24	48			16
20	18	36			12
21	21	42			14
22	45	90			30
23	30	60			20
24	15	30			10
25	48	96			32
26	96	192			64

FIGURE 9.2
The EMLR (British Museum 10250). Length 17½ inches. Courtesy British Museum.

for presuming the existence of standard fraction tables, the G rule, and the two-thirds table.

We first of all rewrite the table of the EMLR to restore the order of difficulty of the equalities, and to systematize the groups into which they most naturally fall.

THE FIRST GROUP

l. 7		$\bar{3}$	$\bar{3} = \bar{\bar{3}}$
l. 5		$\bar{6}$	$\bar{6} = \bar{3}$
l. 4		$\overline{10}$	$\overline{10} = \bar{5}$
l. 6	$\bar{6}$	$\bar{6}$	$\bar{6} = \bar{2}.$

In this first group, the four equalities are so simple* that one might well ask why the scribe wrote them, for the mere understanding of what is meant by a fraction makes their existence self-evident. But they are fundamental, and indeed, just a cursory glance at lines 5 and 6 gives us at once the important relation,

$$\text{A} \qquad \bar{3} \quad \bar{6} = \bar{2},$$

from which so many other equalities are derived. Then, considering lines 7 and 5 together, giving $\bar{3}\ \bar{6}\ \bar{6} = \bar{\bar{3}}$, we find using equation A that

$$\text{B} \qquad \bar{2} \quad \bar{6} = \bar{\bar{3}},$$

an equality of great importance to the scribes.

THE SECOND GROUP

The members of the second group offer themselves at once by their similarity, and we note that in each, the second term is double the first:

l. 11	$\bar{9}$	$\overline{18} = \bar{6}$
l. 13	$\overline{12}$	$\overline{24} = \bar{8}$
l. 24	$\overline{15}$	$\overline{30} = \overline{10}$
l. 20	$\overline{18}$	$\overline{36} = \overline{12}$
l. 21	$\overline{21}$	$\overline{42} = \overline{14}$
l. 19	$\overline{24}$	$\overline{48} = \overline{16}$

* Two fractions that are "alike" are never given by the scribes as data or answers to problems. They are met only in tables and, even then, very rarely.

$$l.\ 23 \quad \overline{30} \quad \overline{60} = \overline{20}$$
$$l.\ 22 \quad \overline{45} \quad \overline{90} = \overline{30}$$
$$l.\ 25 \quad \overline{48} \quad \overline{96} = \overline{32}$$
$$l.\ 26 \quad \overline{96} \quad \overline{192} = \overline{64}.$$

Now, even though these ten equalities are scrambled (so to speak) in the EMLR, the scribe could not have failed to notice that the answer is in every case one-third of the second term. Here is the beginning of the G rule of Chapter 5. Of more importance to him, however, was the recognition of the equality A

$$\overline{3} \quad \overline{6} = \overline{2}.$$

We can note here a further extension of the second group, building on lines 3 and 2 thus:

$$l.\ 3 \quad \overline{4} \quad \overline{12} = \overline{3}$$
$$l.\ 2 \quad \overline{5} \quad \overline{20} = \overline{4}$$
$$\overline{6} \quad \overline{30} = \overline{5}$$
$$\overline{7} \quad \overline{42} = \overline{6}$$
$$\overline{8} \quad \overline{56} = \overline{7}$$

. . . .

This extended table could have been discovered by the scribes, using nothing more than the natural number series for the two outside columns and the continuous addition of the even numbers 8, 10, 12, . . . for the center column.

We shall come back to this table in Chapter 10. For the present, I remark that we have here the faint beginnings of a theory of numbers, which as far as we know remained unwritten for 2,000 years.

THE THIRD GROUP

The third group consists of

$$l.\ 3 \quad \overline{4} \quad \overline{12} = \overline{3}$$
$$l.\ 2 \quad \overline{5} \quad \overline{20} = \overline{4}$$
$$l.\ 1 \quad \overline{10} \quad \overline{40} = \overline{8}.$$

A little deeper thought was required to establish these. But the scribe had more than one method.* First, he can put for line 3

$$\bar{6} \qquad \bar{6} \quad = \bar{3} \qquad \qquad \text{(from line 5)}$$

$$\bar{6} \quad (\overline{12} \quad \overline{12}) = \bar{3} \qquad \qquad \text{(line 5} \times 2)$$

$$(\bar{6} \quad \overline{12}) \quad \overline{12} \; = \bar{3}$$

$$\bar{4} \qquad \qquad \overline{12} \; = \bar{3}. \qquad \text{(equation A} \times 2)$$

Second, he could have used the method of the red auxiliaries described in the preceding chapter. Thus if he wished to add $\bar{4} + \overline{12}$, he could have chosen a reference number, say 12, and then reasoned as follows: applied to 12, $\bar{4}$ is 3, and $\overline{12}$ is 1, so that added together, $\bar{4}$ and $\overline{12}$ is 3 and 1 or 4, which applied to 12 is $\bar{3}$; therefore,

$$\bar{4} \quad \overline{12} = \bar{3}.$$

Third, he could have referred to the previously established equality B, and then, dividing through by 2, he would have $\bar{4} \quad \overline{12} = \bar{3}$.

Of course we cannot know whether the scribe found line 3 by one of these or some other method. But having established it, the successive multiplication by the odd numbers produces the sequence of equalities

$$\overline{12} \quad \overline{36} = \bar{9}$$
$$\overline{20} \quad \overline{60} = \overline{15}$$
$$\overline{28} \quad \overline{84} = \overline{21}$$
$$\cdots$$

Now, by multiplying through by 2 (i.e., halving the denominators), he has

$$\bar{6} \quad \overline{18} = 2 \div \; 9$$
$$\overline{10} \quad \overline{30} = 2 \div 15$$
$$\overline{14} \quad \overline{42} = 2 \div 21$$
$$\cdots$$

* K. Vogel, Vorgriechische Mathematik, Part 1, *Vorgeschichte und Ägypten*, Schroedel, Hanover, 1958. On p. 40 Vogel suggests that from $\frac{3}{4} + \frac{1}{4} = 1$, division by 3 gives $\frac{1}{4} + \frac{1}{12} = \frac{1}{3}$. This may have been the scribe's thought process, but he had no way of writing $\frac{3}{4}$, $\frac{4}{5}$, or $\frac{5}{6}$!

But these are exactly the expressions given in the RMP Recto Table of 2 divided by the odd numbers. Then we are presented with the possibility of deriving line 3 of the EMLR from $(2 \div 3)$ of the Recto, or, vice versa, deriving $(2 \div 3)$ of the Recto from line 3 of the EMLR, which lends some credence, perhaps, to Glanville's thought that the EMLR and the RMP were in some way related.

Line 2, and consequently line 1, can be established in much the same way as line 3:

$$\bar{5} \quad \bar{5} \quad \quad \bar{5} \quad \bar{5} \quad \quad \bar{5} = 1, \qquad \text{(definition of one-fifth)}$$

$$\overline{20} \quad \overline{20} \quad \quad \overline{20} \quad \overline{20} \quad \quad \overline{20} = \bar{4},$$

$$(\overline{20} \quad \overline{20}) \quad (\overline{20} \quad \overline{20}) \quad \overline{20} = \bar{4},$$

$$(\overline{10} \qquad \qquad \overline{10}) \qquad \overline{20} = \bar{4}, \qquad \text{(line 4, } \times 2)$$

$$\bar{5} \qquad \qquad \overline{20} = \bar{4},$$

which is line 2. Doubling each number of this produces line 1. Or, as with the previous groups, the reference number method could be used if the scribe chose.

THE FOURTH GROUP

There are seven 3-term equalities in the EMLR, excluding line 6, of which the five consecutive lines 14–18 are all derivable from the original prototype

$$C \quad \bar{2} \quad \bar{3} \quad \bar{6} = \bar{1},$$

easily established from equality A and $\bar{2}\ \bar{2} = 1$.

$$l.\ 12 \quad \bar{7} \quad \overline{14} \quad \overline{28} = \bar{4}$$

$$l.\ 14 \quad \overline{14} \quad \overline{21} \quad \overline{42} = \bar{7}$$

$$l.\ 15 \quad \overline{18} \quad \overline{27} \quad \overline{54} = \bar{9}$$

$$l.\ 16 \quad \overline{22} \quad \overline{33} \quad \overline{66} = \overline{11}$$

$$l.\ 17 \quad \overline{26} \quad \overline{39} \quad \overline{78} = \overline{13}$$

$$l.\ 18 \quad \overline{30} \quad \overline{45} \quad \overline{90} = \overline{15}$$

$$l.\ 10 \quad \overline{25} \quad \overline{50} \quad \overline{150} = \overline{15}.$$

Lines 14–18 were inscribed in their correct sequence in the EMLR, being derived from equation C, on multiplying the denominators by the odd numbers 7, 9, 11, 13, and 15, respectively. It is this particular sequence that convinces me that the error of line 17 is indeed here properly corrected from

$$\overline{28} \quad \overline{49} \quad \overline{196} = \overline{13},$$

as on the EMLR, to

$$\overline{26} \quad \overline{39} \quad \overline{78} = \overline{13}.$$

It is possible that the scribe deduced the three-term equalities of the fourth group from the entries which we have recorded in the first and second groups. We illustrate how he could have done this for line 18. From the first group, line 4, he has $\overline{10} \ \overline{10} = \overline{5}$, which when multiplied through by 3 gives $\overline{30} \ \overline{30} = \overline{15}$. Then from the second group, line 22, he has $\overline{45} \ \overline{90} = \overline{30}$, giving $\overline{30} \ \overline{45} \ \overline{90} = \overline{15}$ on substitution, which is line 18 of the fourth group. In a similar manner, lines 14–17 may be as easily derived. And, of course, he may have used the method of the red auxiliaries.

THE NUMBER SEVEN

Line 12 of the fourth group,

$$\overline{7} \quad \overline{14} \quad \overline{28} = \overline{4},$$

is perhaps the most interesting of all the 26 entries of the EMLR. To the ancients of all nations, the number 7 seems always to have held some ineffable fascination. We can immediately think of the 7 wonders of the world, the 7 hills upon which Rome was built, the 7 deadly sins, the 7 against Thebes, the 7 champions of Christendom, the 7 muses, devils, sisters, and veils, 7 years' bad luck, and the 7th heaven. The number 7 was constantly before the Egyptian scribe, for in his multiplication, based upon the method of continued doubling (and halving in division), the sequence

$$1 + 2 + 4 = 7$$

appeared frequently. Furthermore, the Egyptian table of length was

$$4 \text{ digits (or fingers)} = 1 \text{ palm,}$$

$$7 \text{ palms} = 1 \text{ royal cubit,}$$

$$\text{hence, } 28 \text{ digits} = 1 \text{ royal cubit.}$$

At once from this the scribe has

$$1 \text{ digit} = \overline{28} \text{ cubit,}$$
$$2 \text{ digits} = \overline{14} \text{ cubit,}$$
$$4 \text{ digits} = \overline{7} \text{ cubit,}$$
$$7 \text{ digits} = \overline{4} \text{ cubit,}$$

so that, since

$$4 + 2 + 1 = 7$$

in digits, then

$$\overline{7} \quad \overline{14} \quad \overline{28} = \overline{4}$$

in cubits.

Establishing this important relationship between unit fractions might very well have been a useful and regular teaching point in the scribal schools, and perhaps its derivation provided a sort of academic pastime for the scribes themselves. Each of the following five derivations illustrates a different method of arriving at the relationship.

1. The scribe in RMP 34 shows that

$$4 \times (1 \quad \overline{2} \quad \overline{4}) = 7;$$

he then writes

$$\overline{7} \text{ of } (1 \quad \overline{2} \quad \overline{4}) = \overline{4},$$

whence $\overline{7} \ \overline{14} \ \overline{28} = \overline{4}$.

2. Since from ordinary multiplication with integers,

$$1 + 2 + 4 = 7,$$

dividing throughout by 7 (using the Recto Table, where, $2 \div 7 = \bar{4}\ \overline{28}$, and hence, $4 \div 7 = \bar{2}\ \overline{14}$) gives

$$\bar{7}(\bar{4} \quad \overline{28})(\bar{2} \quad \overline{14}) = 1;$$

rearranging, this is

$$(\bar{2} \quad \bar{4})(\bar{7} \quad \overline{14} \quad \overline{28}) = 1.$$

Noting that $\bar{2}\ \bar{4}\ \bar{4} = 1$, subtract $\bar{2}\ \bar{4}$ from each side, whence

$$\bar{7}\ \overline{14}\ \overline{28} = \bar{4}.$$

3. Using a reference number (here obviously 28), the reasoning is

$$
\begin{array}{ccc}
\bar{7} & \text{is} & 4, \\
\overline{14} & \text{is} & 2, \\
\overline{28} & \text{is} & 1,
\end{array}
$$

then adding, $\bar{7}\ \overline{14}\ \overline{28}$ is 7, which applied to 28 is $\bar{4}$.

4. The EMLR third group slightly extended is

$$
\begin{array}{ccc}
\bar{4} & \overline{12} & = \bar{3} \\
\bar{5} & \overline{20} & = \bar{4} \\
\bar{6} & \overline{30} & = \bar{5} \\
\bar{7} & \overline{42} & = \bar{6}.
\end{array}
$$

Doubling the last equality gives

$$\overline{14} \quad \overline{84} = \overline{12},$$

and adding these two gives

$$\bar{7} \quad \overline{14} \quad (\overline{42} \quad \overline{84}) = \bar{6} \quad \overline{12};$$

hence $\bar{7}\ \overline{14}\ \overline{28} = \bar{4}$, for both $\overline{42}\ \overline{84} = \overline{28}$ and $\bar{6}\ \overline{12} = \bar{4}$ are members of the second group, in which the second term is always double of the first.

5. The equality can be arrived at by the multiplication of $\bar{4}$ by 7,

$$
\begin{array}{ccc}
\diagdown 1 & \qquad & \diagdown \quad \bar{4} \\
\diagdown 2 & & \diagdown \quad \bar{2} \\
\diagdown 4 & & \diagdown 1 \\
\hline
\text{Totals } 7 & & 1 \quad \bar{2} \quad \bar{4}.
\end{array}
$$

But this means that $1 \ \overline{2} \ \overline{4} = \overline{4} \times 7$; so divide by 7 to get $\overline{7} \ \overline{14} \ \overline{28}$ $= \overline{4}$.

We can be pretty sure that these five examples by no means exhaust the ways in which this important equality could have been arrived at by the scribes.

LINE 10 OF THE FOURTH GROUP

The last equality of the fourth group, line 10 of the EMLR, is an expression quite unlike any of the other 25. It would be most interesting to know just how it was established by the scribe. It is quite easily proved to be true by using the red auxiliaries. Thus, applied to the reference number 150,

$$\overline{25} \quad \text{is} \quad 6,$$
$$\overline{50} \quad \text{is} \quad 3,$$
$$\overline{150} \quad \text{is} \quad 1;$$

then adding, $\overline{25} \ \overline{50} \ \overline{150}$ is 10, which is $\overline{15}$ of 150, so that it is proved that $\overline{25} \ \overline{50} \ \overline{150} = \overline{15}$. It would not be easy to imagine the circumstances in which this equality would arise. None of its terms is prime, and the common factor 5 reduces it to

$$\overline{5} \quad \overline{10} \quad \overline{30} = \overline{3},$$

which the scribe could have established in several different ways, e.g.,

	$\overline{2}$	$\overline{6} =$	$\overline{3}$	(equation B)
1	$\overline{2}$	$\overline{6} = 1$	$\overline{\overline{3}}$	(adding unity)
$\overline{5}$	$\overline{10}$	$\overline{30} =$	$\overline{3}$;	(dividing by 5)

Or perhaps he said: since $6 \ 3 \ 1 = 10$ in integers, divide by 30 to obtain $\overline{5} \ \overline{10} \ \overline{30} = \overline{3}$.

THE FIFTH GROUP

$$1.\ 8 \quad \overline{25} \quad \overline{15}^* \quad \overline{75} \quad \overline{200} = \overline{8}$$
$$1.\ 9 \quad \overline{50} \quad \overline{30}^* \quad \overline{150} \quad \overline{400} = \overline{16}.$$

* This is one of the rare occasions where a set of fractions belonging to the same group is not exactly arranged in descending order of magnitude: $\overline{15}$ should precede $\overline{25}$, and $\overline{30}$ should precede $\overline{50}$.

Using the red auxiliaries, and (as is most usual) taking 200, the greatest denominator for reference number from the left-hand side of the equality of line 8, we would have

$$\overline{25} \quad \text{is} \quad 8,$$
$$\overline{15} \quad \text{is} \quad 13 \quad \overline{3},$$
$$\overline{75} \quad \text{is} \quad 2 \quad \overline{3},$$
$$\overline{200} \quad \text{is} \quad 1;$$

adding, $\overline{25}$ $\overline{15}$ $\overline{75}$ $\overline{200}$ is 25, which is $\overline{8}$ of 200. Therefore line 8 follows. Simple doubling gives line 9.

The above, however, is only a proof, by a standard scribal technique, that the equality is true. It does not show how the four terms were chosen in the first place, nor how the scribe knew their sum to be $\overline{8}$. Perhaps we shall never find out, but we can still hazard a plausible guess as to how he arrived at it. One way, perhaps a most unlikely one, would be for the scribe to have noted that

$$24 \quad 40 \quad 8 \quad 3 = 75,$$

in integers; then, on dividing through by 600, he would have had line 8. But we can establish it more plausibly by selecting $\overline{6}$ $\overline{30}$ $= \overline{5}$ from the extended second group (p. 96). Multiply this through by 5, so that

$$\overline{25} = \overline{30} \quad \overline{150},$$

then add $\overline{15}$ $\overline{75}$ to both sides and regroup, giving

$$\overline{25} \quad \overline{15}* \quad \overline{75} = (\overline{15} \quad \overline{30}) \quad (\overline{75} \quad \overline{150}) \qquad \text{(from}$$
$$= \overline{10} \qquad \overline{50}, \qquad \text{line 24)}$$

then add $\overline{200}$ to obtain

$$\overline{25} \quad \overline{15} \quad \overline{75} \quad \overline{200} = \overline{10} \quad (\overline{50} \quad \overline{200}) \qquad \text{(from line}$$
$$= \overline{10} \qquad \overline{40} \qquad \qquad 2 \times 10)$$
$$= \overline{8}. \qquad \qquad \text{(from line 1)}$$

We can conclude that Vogel, Neugebauer, and Van der Waerden were indeed farseeing when they considered that the EMLR would add considerably to our knowledge and understanding of the Egyptian techniques for handling unit fractions despite Glanville's and Scott and Hall's disenchanted reflections when it was first unrolled.

* Here we see a possible explanation of why $\overline{25}$ preceded $\overline{15}$.

10 UNIT-FRACTION TABLES

In this chapter, most numbers are denominators of unit fractions. Hence the overbars will be omitted unless there is some chance of confusion. The exception to the unit-fraction rule is of course $\frac{2}{3}$, which is written $\bar{\bar{3}}$. In order to identify groups of unit fractions such as

1	2		3	5
2	4	or	6	10
3	6		9	15,

where each fraction is composed of terms that are integral multiples of the terms of the first fraction, we shall refer to each group by its initial pair, calling this pair the generator of the group; i.e., the left-hand group shown here has the generator (1, 2) and the other, (3, 5). Note that, as elsewhere in this book, no plus sign is included in the unit fractions, mere juxtaposition implying addition, just as, say, $4\frac{1}{2}$ means $4 + \frac{1}{2}$ in modern notation. However, of necessity we use the modern sign = for "equals," although the Egyptians had no such sign. In the EMLR there is on each line a symbol, which means "this is." This notation is the nearest counterpart to a modern equals sign in Egyptian arithmetic.

The scribes' preoccupation with operations involving unit fractions rather than integers is quite marked in Egyptian mathematical papyri. Of the 87 problems in the RMP only a mere six do not involve fractions, while the Recto of the RMP, which constitutes nearly one-third of the 18-foot-long papyrus, deals entirely with odd fractional divisions of 2, as we have seen in Chapters 6 and 7. Because the Egyptians performed their multiplications and divisions by doubling and halving, it was necessary to be able to double fractions as well as integers. For fractions with even denominators this was easy; but with odd numbers as denominators this doubling could sometimes be quite difficult, so that a previously prepared table would have been a great aid to computation. Indeed, a portion of the RMP Recto Table is included in the KP. Other papyri, such as the MMP, also show this obsession with unit fractions. Of the 25 problems of the MMP only three do not involve calculations with fractions.

The Egyptians' concern for the accurate dealing with fractions almost certainly originated from practical problems, such as the division of food, supplies, and other things, either equally or in some specific ratio, among families, troops, members of working gangs, large crews, etc., in a country which had no metallic currency or money, and in which payments were made in kind. Only for large payments might weights of precious metals like silver or gold be used (see Chapter 20). If a fractional number like our ⅗ needed to be expressed or evaluated, as would arise for example in the division of 3 loaves among 5 men, the Egyptians had no other way of writing it than as 3̄ 5̄ 15̄ or some other combination of unit fractions, or by saying "three divided by five." Today, if a new concept arises, mathematicians devise at once a new notation for it, but the Egyptians, never thinking to improve or alter their notation for fractions, developed instead special techniques for dealing with the notation they already had. To divide 3 loaves equally among 5 men, each man would be given three separate portions, a ⅓, a ⅕, and a ¹⁄₁₅. One advantage of this division was that not only was justice done, but justice also appeared to have been done. In a modern distribution, three of the five men would get ⅗ of a loaf in one large piece, while the other two men would get two smaller pieces, ⅖ and ⅕ of a loaf, which division might be regarded as an injustice by an ignorant workman.*

The scribe would never write 5̄   5̄   5̄ for ⅗ except in tables used for calculating. No two similar fractions ever occur in the scribe's answer to a problem or a calculation, even though they may have occurred several times repeated in the mechanical operations that the problem required. Just exactly what was the reason at the back of the Egyptian mind for this is still not clear. It is, however, certain that the concepts of "one-half," "one-third," or "one-quarter," and so on, must have come before the invention of the hieroglyphs that denote

* Problem 6 of the RMP requires the division of 9 loaves among 10 men. While the modern answer is that each man gets ⁹⁄₁₀ of a loaf, this division requires that the last man must get (¹⁄₁₀ ¹⁄₁₀ ¹⁄₁₀ ¹⁄₁₀ ¹⁄₁₀ ¹⁄₁₀ ¹⁄₁₀ ¹⁄₁₀ ¹⁄₁₀) of a loaf; already sliced so to speak! The Egyptians would have none of such a solution. The answer given in the RMP is that every man gets exactly the same number of pieces and exactly the same-sized pieces, namely, ⅔, ⅕, and ¹⁄₃₀ each, and justice is again obviously done.

them. Thus also the concepts preceded the conventional notation of writing the fractions with their numbers. And so likewise would the concepts of "the two parts" (in three) and "the three parts" (in four) have come quite early, just as today we often speak of something being "three parts full," when we mean three-quarters full. In modern German, there is a word *anderthalb*, meaning literally another or second half, so that *anderthalb pfund* means $1\frac{1}{2}$ pounds.

If one-fifth, let us say, was cut off from one end of a loaf of bread, the Egyptian did not seem to think of what was left as being $\frac{4}{5}$ of the original loaf, but rather as the loaf reduced; if a fifth of this reduced loaf was cut off, the two "fifths" thus removed would of course be unequal. Perhaps this circumstance may help us to understand why the scribes so carefully avoided writing $\bar{5}\ \bar{5}$ for $\frac{2}{5}$ and wrote instead $\bar{3}\ \overline{15}$. The ancient Egyptians might even have claimed with some justification that their method of division was superior!

Divisions of loaves and other commodities could often be further complicated by the provisos that overseers were to get three times as much as the boatmen and doorkeepers, who in turn were to get double portions. How will the supervisor cut up the loaves if each man expects to get the same number of pieces as other workers of his class? There were many such problems of partitioning, some far more complicated than this, and thus inevitably, when the system of writing numbers and fractions had been evolved, one of the first requirements was to prepare useful tables of unit-fractional equivalents for computation. The EMLR is one such table, though perhaps a rather sophisticated one. In the RMP there are, besides the Recto Table, five other shorter tables of equivalent fractions. Other similar tables are to be found in other papyri, on ostraca, on leather, even on wood.

UNIT-FRACTION TABLES OF THE RHIND MATHEMATICAL PAPYRUS

The Recto of the RMP, as we have already seen, contains decompositions of the division of 2 by the odd numbers 3 to 101 (see Chapter 7). In each division, the scribe first gave his answer, and then proved it correct by multiplication of the fractions involved in each answer, which necessitated the addition of several fractions to total 2. In only one of the 50 divisions does the scribe show how he "knows" that the fractions concerned do in fact total 2 (see pp. 77–79), but they all do, and nowhere has he made an error. A close examination of the Recto leads one to the conclusion that the scribe must have referred to some set of tables giving the sums of 2-term and 3-term unit fractions. Either that, or he worked each one out separately on a papyritic memo pad, or he was superbly competent at mental arithmetic. The first conclusion is the most likely, for it is already well established that the Egyptians made constant use of various types of tables. The set of tables that the scribe who was calculating the divisions of the Recto had for reference must have contained the equalities given in Tables 10.1 and 10.2. Whether the simple relations $\bar{2}\ \bar{2} = 1$ and $\bar{4}\ \bar{4} = \bar{2}$ were included or not is a matter of conjecture. They occur frequently in the problems (17 times in the Recto and more than 40 times in the Verso); they might very well have been considered to be too simple to bother recording in a standard table. The same would apply to the relation $\bar{3}\ \bar{3} = 1$, which appears in at least 14 of the Verso problems. In this list, each equality is easily located either in the Recto or the Verso of the RMP as shown, but I am sure that Tables 10.1 and 10.2 are by no means complete. That these tables are not as complete as the scribes' may have been can be illustrated by the last entry for the gen. (4, 5, 20). It is possible that the scribe first noted that $(10\ 40) = 8$ from gen. (1, 4) and then put $(8\ 8) = 4$ from gen. (1, 1), but we are unable to decide precisely which equality was used. Not only were all the equalities in these tables used by the scribe as a commonplace procedure not requiring any proof or justification: he sometimes went much further; for example, in Problem 65, he wrote without comment $100 \div 13 = 7\ \bar{3}\ \overline{39}$, as if he did the division mentally, which may well have been the case. But to say the same of the two following equalities, particularly the last, would be to strain our credulity too far!

$$\bar{2} \quad \bar{5} \quad \overline{10} \quad \overline{10} \quad \overline{10} = 1, \qquad \text{(Problem 35)}$$

$$\bar{2}\ \bar{6}\ \overline{12}\ \overline{14}\ \overline{21}\ \overline{21}\ \overline{42}\ \overline{63}\ \overline{84}\ \overline{126}\ \overline{126}\ \overline{168}\ \overline{252}$$
$$\overline{336}\ \overline{504}\ \overline{1008} = 1. \quad \text{(Problem 70)}$$

If the case for the use of tables has not been completely established up to this point, then surely this last equality would clinch it. The reader

TABLE 10.1
Two-term equalities appearing in the RMP.

Generator	Equality			Occurrence
(1, 1)	6	6 =	3	Recto, 2 ÷ 11, 35, 41, 53, 55.
				Verso, Problem 40.
	8	8 =	4	Recto, 2 ÷ 13, 67, 71, 97.
	10	10 =	5	Recto, 2 ÷ 89.
	14	14 =	7	Verso, Problem 69.
(1, 2)	3	6 =	2	Recto, 2 ÷ 17, 37, 47, 73, 79, 83, 89, 95.
				Verso, Problem 69.
	6	12 =	4	Recto, 2 ÷ 19, 23.
	21	42 =	14	Verso, Problem 69.
	81	162 =	54	Verso, Problem 42.
(1, 3)	4	12 =	3	Recto, 2 ÷ 95.
	8	24 =	6	Recto, 2 ÷ 29, 37, 41.
(1, 4)	5	20 =	4	Recto, 2 + 31, 67, 73, 83, 89.
	10	40 =	8	Recto, 2 ÷ 71.
(1, 6)	7	42 =	6	Recto, 2 ÷ 43. Verso, Problem 69.
(1, 7)	8	56 =	7	Verso, Problems 7, 7B, 11.
	16	112 =	14	Verso, Problems 7, 7B, 11.
(2, 3)	10	15 =	6	Recto, 2 ÷ 47, 53, 79.
	20	30 =	12	Verso, Problem 76.

TABLE 10.2
Three-term equalities appearing in the RMP.

Generator	Equality				Occurrence
(1, 1, 1)	6	6	6 =	2	Recto, 2 ÷ 29.
(1, 2, 4)	7	14	28 =	4	Recto, 2 ÷ 97. Verso, Problems 38, 69.
	14	28	56 =	8	Verso, Problem 24.
(1, 2, 6)	5	10	30 =	3	Recto, 2 ÷ 91. Verso, Problems 1, 3, 6, 30.
(1, 3, 3)	5	15	15 =	3	Verso, Problems 2, 30.
(1, 3, 5)	23	69	115 =	15	Verso, Problem 30.
	46	138	230 =	30	Verso, Problem 30.
(2, 3, 6)	2	3	6 =	1	Recto, 2 ÷ 43, 101. Verso, Problems 66, 69.
	6	9	18 =	3	Verso, Problems 17, 42, 67.
	18	27	54 =	9	Verso, Problem 42.
	36	54	108 =	18	Verso, Problem 42.
(3, 4, 6)	9	12	18 =	4	Recto, 2 ÷ 59.
(3, 10, 15)	3	10	15 =	2	Verso, Problems 4, 5, 35, 56.
(4, 5, 20)	8	10	40 =	4	Recto, 2 ÷ 61.

is invited to check this equality by modern arithmetic, or by using the scribes' methods.

PROBLEMS 7 TO 20 OF THE RHIND MATHEMATICAL PAPYRUS
These 15 problems (including Problem 7B) form in reality a table of 3- and 6-term unit-fraction equalities, similar to the Recto Table and the tables of the EMLR. The two numbers 1 $\overline{2}$ $\overline{4}$ and 1 $\overline{3}$ $\overline{3}$ are used as multipliers for a succession of multiplicands, so chosen as to result in certain equalities likely to be of use in the later problems of the papyrus. Chace regards them as examples of simple multiplications of fractional expressions, and Neugebauer as completion problems for 2 ÷ 7 and 2 ÷ 9 of the Recto. The following multiplications were carried out by the scribe:

1	$\overline{2}$	$\overline{4}$ times	$\overline{7}$		(Pr. 11)
		,,	$\overline{14}$		(Pr. 12)
		,,	$\overline{28}$		(Pr. 14)
		,,	$\overline{2}$	$\overline{14}$	(Pr. 9)
		,,	$\overline{4}$	$\overline{28}$	(Prs. 7, 7B, 10)
		,,	$\overline{16}$	$\overline{112}$	(Pr. 13)
		,,	$\overline{32}$	$\overline{224}$,	(Pr. 15)
1	$\overline{3}$	$\overline{3}$ times	$\overline{2}$		(Pr. 16)
		,,	$\overline{3}$		(Pr. 17)
		,,	$\overline{4}$		(Pr. 8)
		,,	$\overline{6}$		(Pr. 18)
		,,	$\overline{12}$		(Pr. 19)
		,,	$\overline{24}$.		(Pr. 20)

As a result of these 15 multiplications, the following equalities are established in this group of problems:

$\overline{7}$	$\overline{14}$	$\overline{28}$ =	$\overline{4}$	(Prs. 11, 10)
14	28	56 =	8	(Prs. 12, 9)
28	56	112 =	16	(Pr. 14)
112	224	448 =	64	(Pr. 13)
224	448	896 =	128,	(Pr. 15)

$$\bar{2} \quad \bar{3} \quad \bar{6} = \bar{1} \qquad \text{(Pr. 16)}$$
$$\bar{4} \quad \bar{6} \quad \overline{12} = \bar{2} \qquad \text{(Pr. 8)}$$
$$\bar{6} \quad \bar{9} \quad \overline{18} = \bar{3} \qquad \text{(Prs. 18, 17)}$$
$$\overline{12} \quad \overline{18} \quad \overline{36} = \bar{6} \qquad \text{(Pr. 19)}$$
$$\overline{24} \quad \overline{36} \quad \overline{72} = \overline{12}. \qquad \text{(Pr. 20)}$$

Then, from

$$1 = \bar{2} \quad \bar{2}$$
$$= \bar{2} \quad \bar{4} \quad \bar{4}$$
$$= \bar{2} \quad \bar{4} \quad \bar{8} \quad \bar{8},$$

and since $\overline{14}$ $\overline{28}$ $\overline{56} = \bar{8}$ from the foregoing list,

	$\bar{2}$	$\bar{4}$	$\bar{8}$	$\overline{14}$	$\overline{28}$	$\overline{56}$ = 1	(Pr. 9)
×2	$\bar{4}$	$\bar{8}$	$\overline{16}$	$\overline{28}$	$\overline{56}$	$\overline{112}$ = 2	(Prs. 10, 7, 7B)
×4	$\overline{16}$	$\overline{32}$	$\overline{64}$	$\overline{112}$	$\overline{224}$	$\overline{448}$ = 8	(Pr. 13)
×2	$\overline{32}$	$\overline{64}$	$\overline{128}$	$\overline{224}$	$\overline{448}$	$\overline{896}$ = 16.	(Pr. 15)

It is my view that this group of problems either was included in the RMP to establish a set of 3-term and 6-term equalities for inclusion in Egyptian standard tables, or was taken from a set of such tables.

Working only with the same methods, techniques, and notations available to an Egyptian scribe, we now attempt to reproduce some of these unit-fraction tables ab initio. We must of course eschew any modern refinements that could lead us to obvious simplifications. It may be a little irksome, but we have to try to think how the scribe would have thought, to imagine we are writing in hieratic, and to be logical only to the extent that we could expect the scribe to have been logical. And so our very first table of fractions will come from the original meaning of the word "fraction," a part, where two halves, three thirds, four quarters, and so on, make one whole. Then we have our first and most elementary table (Table 10.3). Each of the unit-fraction equalities of Table 10.3 will produce other equalities by the use of generators, i.e., by the successive multiplication by integers. Table 10.4a thus appears by so treating the first equality of Table 10.3. To Table 10.4a could be added its only odd-number member, $\bar{3}$ $\bar{3}$ = $\bar{\bar{3}}$, and that only because of the uniqueness of $\bar{3}$. Similarly from the other equalities of the basic Table 10.3, Tables 10.4b–d arise.

TABLE 10.3
The most elementary table, from the definition of a fraction. Overbars not shown.

			2	2 = 1
		3	3	3 = 1
	4	4	4	4 = 1

. . .

TABLE 10.4
The table produced by applying generators to the equalities of Table 10.3. The equality $\overline{3}$ $\overline{3}$ = $\overline{\overline{3}}$ can be considered to be included in a. Overbars omitted from this table.

Generator (1, 1)	Generator (1, 1, 1)	Generators (1, 1, 1, 1)	Generators (1, 1, 1, 1, 1)
2 2 = 1	3 3 3 = 1	4 4 4 4 = 1	5 5 5 5 5 = 1
4 4 = 2	6 6 6 = 2	8 8 8 8 = 2	10 10 10 10 10 = 2
6 6 = 3	9 9 9 = 3	12 12 12 12 = 3	15 15 15 15 15 = 3
. . .	. . .	. . .	. . .
(a)	(b)	(c)	(d)

If Table 10.4 appears to be elementary, it is worth noting that the scribe of the EMLR thought some of the entries worthy of being recorded in that work:

line 4 of the EMLR,		10	10 = 5;
line 5 of the EMLR,		6	6 = 3;
line 6 of the EMLR,	6	6	6 = 2;
line 7 of the EMLR,		3	3 = $\overline{\overline{3}}$.

This suggests that lists similar to those in Table 10.4 were at some time constructed by the scribes, and were in fact used by them, even though no such tables have come down to us in toto. Their most probable use would have been in the scribal schools.

We now endeavor to establish further equalities derivable by the scribes from the Tables 10.3 and 10.4.

We begin with Table 10.4a. Adding 6 to each side of line 3 gives

$$3 \quad 6 = 6 \quad 6 \quad 6.$$

The right-hand side of this equals 2 by the generator (1, 1, 1) (line 2 of Table 10.4*b*), or

$$3 \quad 6 = 2.$$

This equality could have been derived otherwise by the scribes. For example, from generator (1, 1) (Table 10.4*a*) they could have written 3 3 = $\bar{3}$ and 6 6 = 3; then by addition,

$$(3 \quad 6) \quad (3 \quad 6) = 1,$$

or, by applying line 1 of generator (1, 1), 3 6 = 2.

Now the equality 3 6 = 2 belongs to a new group, for since the left-hand side can be written as (3 × 1 3 × 2), its generator is (1, 2); successive multiplication of this generator by 2, 3, 4, ... , produces the table

$$3 \quad 6 = 2$$
$$6 \quad 12 = 4$$
$$9 \quad 18 = 6$$

. . . .

The next table to look for is, naturally, that whose generator is (1, 3), and then, if successful, (1, 4), (1, 5), and so on.

Adding 12 to each side of line 2 of the just-generated table gives

$$4 \quad 12 = 6 \quad 12 \quad 12$$
$$= 6 \quad \quad 6 \quad \text{(from gen. (1, 1))}$$
$$= \quad 3. \quad \text{(from gen. (1, 1))}$$

Again, this equality may be otherwise derived. Thus, from generators (1, 2) and (1, 1) we have

$$3 \quad 6 \quad = 2$$
$$(6 \quad 6) \quad (12 \quad 12) = 2$$
$$6 \quad 6 \quad 12 \quad 12 = 3 \quad 6$$
$$6 \quad 12 \quad 12 = 3$$
$$4 \quad 12 = 3.$$

Successive multiplication of this last equality produces the table of the generator $(1, 3)$:

$$
\begin{array}{rrr}
4 & 12 = 3 \\
8 & 24 = 6 \\
12 & 36 = 9
\end{array}
$$

. . . .

We proceed:

gen. $(1, 1)$		5 =	10	10		
gen. $(1, 1)$		=	(20	20)	(20	20)
Add 20 to each side	5	20 =	(20	20	20	20	20)
gen. $(1, 1, 1, 1, 1)$ gives	5	20 =		4.		

Then the table based on generator $(1, 4)$ follows:

$$
\begin{array}{rrr}
5 & 20 = & 4 \\
10 & 40 = & 8 \\
15 & 60 = & 12
\end{array}
$$

. . . .

Again,

gen. $(1, 2)$	15		30	= 10
gen. $(1, 1)$	15	(60	60)	= 10
	(15	60)	60	= 10
gen. $(1, 4)$	12		60	= 10
Divide both sides by 2	6		30	= 5.

We have then the table of generator $(1, 5)$:

$$
\begin{array}{rrr}
6 & 30 = & 5 \\
12 & 60 = & 10 \\
18 & 90 = & 15
\end{array}
$$

. . . .

The scribes no doubt also noted that, besides multiplying the preceding equalities to produce new equalities, they could also add or

TABLE 10.5
Ordered table of generators derived solely from the simpler generators of Table 10.4. The initial equality associated with each generator is given opposite the generator itself.

gen.			
(1, 1)	2	2	= 1
(1, 2)	3	6	= 2
(1, 3)	4	12	= 3
(1, 4)	5	20	= 4
(1, 5)	6	30	= 5
. . .		. . .	

even subtract them to obtain new equalities, provided that those so operated on belonged to the same generator group.

At this stage, the scribes might well have looked over what they so far had achieved and then set down in order the equalities established, making a table of the basic generators (Table 10.5). We may assume that it was not difficult for the scribes to see that they had lighted upon a simple property of sequences of fractions. I suggest that such a discovery by the scribes, and the embryonic theory of numbers that it implies, might help to explain why so many of the problems of the mathematical papyri are freely interspersed with illustrations worked out in fractions when integers would have sufficed. By a simple induction the scribes could have concluded that the series of natural numbers in the table of generators (Table 10.5) could be extended as far as they pleased. The first and third columns of numbers consist simply of the natural number series, and the middle column progresses in a uniform way, for their differences in order are the even numbers 4, 6, 8, ... More obvious perhaps would be the observation that in each equality, the product of the two outside numbers equals the middle one. And further, each element of the table of generators produces another unending table upon the successive multiplication by 2, 3, 4, . . . , so that a comprehensive set of tables could be constructed, containing all possible 2-term unit-fraction equivalents. The EMLR is probably one small selection from such a set of tables.

However powerful the 2-term equalities may have been as an aid to computation, there was no need to stop at 2, and indeed the scribes did not, as the EMLR itself illustrates. Three-term equalities may be derived, sometimes very easily, from the 2-term equality tables already established, and this we now examine. The first and most obvious 3-term equality to establish has the generator (2, 3, 6), because it arises from the first of the 2-term equalities:

gen. (1, 1)			2	2 = 1	
gen. (1, 2)				2 = 3	6
Substitution		2	3	6 = 1.	

For the Egyptian scribe, this equality was almost a standard form; it often occurs in calculations. In like manner the scribe could write

gen. (1, 2)			3	6 = 2	
gen. (1, 3)				6 = 8	24
Substitution		3	8	24 = 2.	

This is an entirely new 3-term equality. The field would begin to widen at this stage, because there happen to be other substitutions available for 6, which the scribe can make; thus

gen. (1, 2)			3	6 = 2	
gen. (1, 2)				6 = 9	18
Substitution		3	9	18 = 2.	

The three equalities deduced so far turn out to be *prime* equalities; i.e., they cannot be simplified by dividing their terms through by some common factor. But, as previously, each may be multiplied in succession by the integers to produce further endless sets of tables. Looking at the last two of these 3-term equalities, we observe, as no doubt the Egyptian scribes did, that since 3 8 24 = 2 and 3 9 18 = 2, then surely

$$8 \quad 24 = 9 \quad 18.$$

Such equalities, composed of members both of which are themselves multi-termed, may have been another source of thought for the ancient Egyptian student of numbers. In the Recto of the RMP, where the scribe divided 2 by 17, we find 3 12 replaced by the more convenient 4 6 without any explanation. And in dividing 2 by 23 he wrote $\overline{3}$ 4 for 2 4 6, again without explanation. The reason for the substitution would appear to be because the 2-term equivalent is shorter, and also perhaps because it contains his favored $\overline{3}$. So equalities composed of multi-term members may have been another facet in his vista of unit-fraction equalities.

Let us now suppose that, in endeavoring to establish a certain 3-term equality, it became necessary to refer to the 2-term tables for the single-unit-fraction sum 28 70. Now this pair of fractions should be found in the table whose generator is (2, 5). But we do not have such a table, and in fact at this stage we do not know whether such a sum has a single-unit-fraction equivalent. Therefore we must consider such pairs before proceeding with the tables for 3-term equalities. We must determine if they in fact do exist, and, if they do, find out how to develop them in a properly ordered sequence.

Although the scribes may already have noted how the table of generators could be extended merely by extending certain series, there remains the question of the relations between the three individual members of each equality. We look again at the table of generators (Table 10.5) that has already been established. There are many properties which an observant student of these equalities might discern, but the relevant ones are those which apply in exactly the same way to all five of them, as well as to all the succeeding equalities. One such property may be seen by taking the last one listed as an example: if the *elements* of the generator of (1, 5) be added, then 6 results; and if the elements be then multiplied by this 6, one obtains the left-hand side of the equality, namely, 6 30. Clearly this simple property is possessed by all the other generators and their corresponding equalities of Table 10.5. The right-hand side of each equality is equal to the second element of the corresponding generator, which in this case is the same as the product of the two elements. Reasonably enough, the scribe might have wondered if such simple relations would apply to other generators when the first element is other than unity. The following are a likely set of generators he would set down to test this conjecture.

(2, 1)	(3, 1)	(4, 1)	(5, 1)
(2, 2)	(3, 2)	(4, 2)	(5, 2)
(2, 3)	(3, 3)	(4, 3)	(5, 3)
(2, 4)	(3, 4)	(4, 4)	(5, 4)
(2, 5)	(3, 5)	(4, 5)	(5, 5)

. . . .

In this array there are some duplications. For example, (2, 2) is the same generator as (1, 1), (2, 4) is the same as (1, 2), and (3, 1) is the same as (1, 3), and so on. To eliminate duplications, we may exclude those pairs of generators having a common factor. Then the generators to be examined are those in the amended array:

(2, 3)	(3, 4)	(4, 5)	(5, 6)
(2, 5)	(3, 5)	(4, 7)	(5, 7)
(2, 7)	(3, 7)	(4, 9)	(5, 8)
(2, 9)	(3, 8)	(4, 11)	(5, 9)
(2, 11)	(3, 10)	(4, 13)	(5, 11)

. . . .

Applying the rules deduced above to the first pair of generators in this array, the scribe would have: the sum of the elements is 5, and multiplying the elements 2 and 3 by this 5 gives 10 15, which should be the left-hand side of the resulting equality. For the right-hand side he should have either the second member 3, or the product $2 \times 3 = 6$. Now, according to the development that I am postulating here, the scribe would have known at once that the sum cannot be 3, for he would remember generator (1, 1), line 3 (Table 10.4*a*), i.e., in which 6 6 = 3 and both $\overline{10}$ and $\overline{15}$ are smaller fractions than $\overline{6}$. The scribe then had to test the truth of the putative equality

$$10 \quad 15 = 6. \qquad \qquad \text{gen. (2, 3)}$$

If the equality is not true, he need have gone no further. If it is true, he would still have had to try several more in order to justify a generalization. The proof follows on the next page.

From gen. (1, 5)		10	=	12	60	
Add 15 to each side	10	15	=	12	(15	60)
From gen. (1, 4)			=	12		12
From gen. (1, 1)	10	15	=		6.	

A proof such as this would have established the truth of this equality for the scribe. He would next try other equalities. For example, he might have taken the generator (3, 4). Following the same methodology as before, the relevant numbers are $3 + 4 = 7$ and $3 \times 4 = 12$, so that the resulting equality which had to be tested is 21 28 = 12 (gen. (3, 4)):

From gen. (1, 7)			8	56	=	7		
Add 42 to each side	8	42	56	=	7		42	
From gen. (1, 6)				=		6		
From gen. (1, 3)				=	8		24	
Then,	8	42	56	=	8		24	
Subtract 8 from each side		42	56	=		24		
Divide by 2		21	28	=		12.		

These two equalities having been established, the scribe would have treated a third in the same way, from the generator (4, 5), say; since here $4 + 5 = 9$ and $4 \times 5 = 20$, the equality to be tested is 36 45 = 20 (gen. (4, 5)):

From gen. (1, 9)			10	90	=	9		
Add 72 to each side	10	72	90	=	9		72	
From gen. (1, 8)				=		8		
From gen. (1, 4)				=	10		40	
Then	10	72	90	=	10		40	
Subtract 10 from each side		72	90	=		40		
Divide by 2		36	45	=		20.		

There is a sameness about each of the three preceding proofs, which suggests that, given any two numbers at all as elements of a generator, the same approach will establish the relevant 2-term equality. However, by choosing the numbers 5 and 7 for elements, the scribe would have found that the method does not always work, in this case because 5 and 7 are both odd numbers. So he would have had to look for another method, which he easily found. For example, the scribe could have established 60 84 = 35 (gen. (5, 7)):

From gen. (1, 6)		60 = 70	420	
Add 84 to each side	60	84 = 70	(84	420)
From gen. (1, 5)		= 70	70	
From gen. (1, 1)	60	84 =	35.	

Trying the same procedure with the generator (3, 5) gives

From gen. (1, 4)		24 = 30	120	
Add 40 to each side	24	40 = 30	(40	120)
From gen. (1, 3)		= 30	30	
From gen. (1, 1)	24	40 =	15.	

If the scribe had tried as generator (3, 7), his technique would have had to be slightly different (he would have had to start with the larger of the summed unit fractions). Thus, to establish 30 70 = 21 (gen. (3, 7)):

From gen. (1, 5)		70 = 84	420	
Add 30 to each side	30	70 = 84	(30	420)
From gen. (1, 14)		= 84	28	
From gen. (1, 3)	30	70 =	21.	

Again, the scribe interested in what we have called his "theory of numbers" or, more properly, his "theory of fractions" would naturally enough stop at about this stage and take stock, in an orderly fashion, of what he had so far established. A collection of 2-term equalities obtainable by means of the methods just discussed is shown in Table A11.1. The established 2-term equalities, together with their generators, are in roman type; an overall view of such a collection would have at once revealed to the scribe many ordered sequences, what today we would call *series*. Indeed, such is the regularity of the various columns that it would be a very unimaginative scribe indeed who would not be tempted to extend his table upwards, as shown by the italic entries (this is downward, so to speak, in the magnitudes of the denominators). By so doing, the scribe would see that he has really rewritten his table so that entries in corresponding positions above and below the diagonal are the same; thus the table reads the same either horizontally or vertically (see Appendix 11).

11 PROBLEMS OF EQUITABLE DISTRIBUTION AND ACCURATE MEASUREMENT

DIVISION OF THE NUMBERS 1 TO 9 BY 10

TABLE 11.1
Quotients of 1, 2, . . . , 9 divided by 10, as listed in the RMP.

Number	Quotient		
1	$\overline{10}$		
2	$\overline{5}$		
3	$\overline{5}$	$\overline{10}$	
4	$\overline{3}$	$\overline{15}$	
5	$\overline{2}$		
6	$\overline{2}$	$\overline{10}$	
7	$\overline{\overline{3}}$	$\overline{30}$	
8	$\overline{\overline{3}}$	$\overline{10}$	$\overline{30}$
9	$\overline{\overline{3}}$	$\overline{5}$	$\overline{30}$

Table 11.1 is a translation of the one made by the scribe in preparation for the first six problems of the RMP. In these problems, which immediately follow the original table, 1, 2, 6, 7, 8, and 9 loaves are to be divided equally among 10 men. In each case the answers are given in the form of those unit fractions shown in the table. In each problem the scribe first states his answer from the table, and then by multiplication he proves that it is correct. We are here more interested in how the table was prepared rather than in its application to the division of loaves among workmen, and we now attempt to reconstruct the scribe's original derivation of this table. The following calculations show how the table could have been derived by straightforward division.

$$
\begin{array}{cc}
1 & 10 \\
\backslash\overline{10} & \backslash\,1 \\
\hline
\overline{10} & 1.
\end{array}
\qquad
\begin{array}{cc}
\dfrac{1}{\overline{10}} & \dfrac{10}{1} \\
\backslash\,\overline{5} & \backslash\,2 \\
\hline
\overline{5} & 2.
\end{array}
$$

$$
\begin{array}{ll}
1 & 10 \\
\diagdown\,\overline{10} & \diagdown\ 1 \\
\diagdown\ \overline{5} & \diagdown\ 2 \\
\hline
\overline{5}\ \overline{10} \qquad 3. & \qquad\quad 3.
\end{array}
\qquad\qquad
\begin{array}{ll}
1 & 10 \\
\diagdown\ \overline{\overline{3}} & \diagdown\ 6\ \overline{\overline{3}} \\
\diagdown\,\overline{30} & \diagdown\ \ \overline{3} \\
\hline
\overline{3}\ \overline{30} \qquad 7. & \qquad\quad 7.
\end{array}
$$

$$
\begin{array}{ll}
1 & 10 \\
\overline{\overline{3}} & 6\ \overline{\overline{3}} \\
\diagdown\ \overline{3} & \diagdown\ 3\ \overline{3} \\
\diagdown\,\overline{15} & \diagdown\ \ \overline{3} \\
\hline
\overline{3}\ \overline{15} \qquad 4. & \qquad\quad 4.
\end{array}
\qquad\qquad
\begin{array}{ll}
1 & 10 \\
\diagdown\ \overline{\overline{3}} & \diagdown\ 6\ \overline{\overline{3}} \\
\diagdown\,\overline{10} & \diagdown\ 1 \\
\diagdown\,\overline{30} & \diagdown\ \ \overline{3} \\
\hline
\overline{3}\ \overline{10}\ \overline{30} \qquad 8. & \qquad\quad 8.
\end{array}
$$

$$
\begin{array}{ll}
1 & 10 \\
\diagdown\,\overline{2} & \diagdown\ 5 \\
\hline
\overline{2} \qquad 5. & \qquad\quad 5.
\end{array}
\qquad\qquad
\begin{array}{ll}
1 & 10 \\
\diagdown\ \overline{\overline{3}} & \diagdown\ 6\ \overline{\overline{3}} \\
\diagdown\ \overline{5} & \diagdown\ 2 \\
\diagdown\,\overline{30} & \diagdown\ \ \overline{3} \\
\hline
\overline{3}\ \overline{5}\ \overline{30} \qquad 9. & \qquad\quad 9.
\end{array}
$$

$$
\begin{array}{ll}
1 & 10 \\
\diagdown\ \overline{2} & \diagdown\ 5 \\
\diagdown\,\overline{10} & \diagdown\ 1 \\
\hline
\overline{2}\ \overline{10} \qquad 6. & \qquad\quad 6.
\end{array}
$$

But the table could have been otherwise constructed. The alternate technique of the succeeding paragraphs may appear to be much longer, but this is only because of the explanatory matter. Actually it is a good deal shorter than the straightforward division just shown. It was possible to write at once the division of 1, 2, and 5 by 10:

Number	Quotient
1	$\overline{10}$
2	$\overline{5}$
5	$\overline{2}$.

From these entries, the scribe could immediately obtain $3 \div 10$ from $(1 + 2) \div 10$, so that he had his third entry $\overline{5}\ \overline{10}$. In like manner, since $6 \div 10$ is $(1 + 5) \div 10$, the sixth entry is at once $\overline{2}\ \overline{10}$. The scribe could also have found $6 \div 10$ by doubling $3 \div 10$, but most probably rejected it as less simple. Performing the doubling, however, gives the unit-fraction decomposition

$$2 \times (3 \div 10) = (2 \div 5) + (2 \div 10)$$
$$= (\bar{3} \quad \overline{15}) \quad \bar{5},$$

where the $\bar{3}$ $\overline{15}$ comes from the Recto Table (Table 6.1). Thus the scribe would have obtained the useful equality $\bar{3}$ $\bar{5}$ $\overline{15} = \bar{2}$ $\overline{10}$. The fourth entry can come from $1 + 3$ or from 2×2, but in each case the scribe would have had $\bar{5}$ $\bar{5}$, which was not acceptable, the two fractions being the same (Precept 3, p. 49). But since two-fifths may be expressed as $2 \div 5$, he had only to look in the Recto Table to find the quotient $\bar{3}$ $\overline{15}$. He now had

Number	Quotient	
1	$\overline{10}$	
2	$\bar{5}$	
3	$\bar{5}$	$\overline{10}$
4	$\bar{3}$	$\overline{15}$
5	$\bar{2}$	
6	$\bar{2}$	$\overline{10}$
7		
8		
9		

The scribe had available several alternate ways of obtaining the quotients for 7, 8, and 9. Thus, he could have found $7 \div 10$ by considering

$$(1 + 6) \div 10 = \overline{10} \quad (\bar{2} \quad \overline{10}) \qquad = \bar{2} \quad \bar{5}.$$
$$(2 + 5) \div 10 = \bar{5} \quad \bar{2} \qquad\qquad = \bar{2} \quad \bar{5}.$$
$$(3 + 4) \div 10 = (\bar{5} \quad \overline{10})(\bar{3} \quad \overline{15}) = \bar{3} \quad \bar{5} \quad \overline{10} \quad \overline{15}.$$

But he would have none of these, because now he could include his most important (and his largest) fraction $\bar{3}$, which was not possible for the preceding dividends. In the same way for $8 \div 10$, he had

$$(1 + 7) \div 10 = \overline{10} \quad ? \qquad\qquad = \quad ?$$
$$(2 + 6) \div 10 = \bar{5} \quad (\bar{2} \quad \overline{10}) \qquad = \bar{2} \quad \bar{5} \quad \overline{10}.$$
$$(3 + 5) \div 10 = (\bar{5} \quad \overline{10}) \quad \bar{2} \qquad = \bar{2} \quad \bar{5} \quad \overline{10}.$$
$$(4 + 4) \div 10 = (\bar{3} \quad \overline{15})(\bar{3} \quad \overline{15}) = (\bar{3} \quad \bar{3}) \quad (\overline{15} \quad \overline{15})$$
$$= \bar{3} \quad \overline{10} \quad \overline{30}.{*}$$

* From $2 \div 15 = \overline{10}$ $\overline{30}$, RMP Recto Table.

This last value is the one the scribe chose, most probably because of the presence of the $\overline{\overline{3}}$.

The table was now complete except for 7 and 9.

Number	Quotient		
1	$\overline{10}$		
2	$\overline{5}$		
3	$\overline{5}$	$\overline{10}$	
4	$\overline{3}$	$\overline{15}$	
5	$\overline{2}$		
6	$\overline{2}$	$\overline{10}$	
7			
8	$\overline{\overline{3}}$	$\overline{10}$	$\overline{30}$
9			

These last two now could have been easily found from $9 \div 10 = (8 + 1) \div 10$ (adding $\overline{10}$ to $\overline{\overline{3}}$ $\overline{10}$ $\overline{30}$ to give $\overline{\overline{3}}$ $\overline{5}$ $\overline{30}$), and from $7 \div 10 = (8 - 1) \div 10$ (subtracting 10 from $\overline{\overline{3}}$ $\overline{10}$ $\overline{30}$ to give $\overline{\overline{3}}$ $\overline{30}$).

CUTTING UP OF LOAVES

The scribe now had the complete list of divisors and quotients as shown in Table 11.1. As I have stated, no working is shown in the RMP for the nine equalities of Table 11.1. To show their uses the scribe chose six practical problems: the division of 1, 2, 6, 7, 8, and 9 loaves among 10 men. The scribe read off the answers from the table; he proved that his choices were correct by performing the appropriate multiplication by 10, as in the following translation of Problem 3:

Divide 6 loaves among 10 men.
Make thou the multiplication $\overline{2}$ $\overline{10}$ times 10.

The doing as it occurs.

1		$\overline{2}$	$\overline{10}$		
\2	\1	$\overline{5}$			
4	2	$\overline{3}$	$\overline{15}$		$[2 \div 5 = \overline{3}$ $\overline{15}$, Recto Table]
\8	\4	$\overline{3}$	$\overline{10}$	$\overline{30}$	$[2 \div 15 = \overline{10}$ $\overline{30}$, Recto Table]

Total 10	6 the same, this is.	$[\overline{5}$ $\overline{10}$ $\overline{30}$ = $\overline{3}$, Table 10.2]

The calculations included by the scribe for divisions of 1, 2, 7, 8, and 9 loaves are similar to this. These calculations are reproduced here:

1	$\overline{10}$				1	$\overline{5}$		
\2	\ $\overline{5}$				\2	\ $\overline{3}$ $\overline{15}$		
4	$\overline{3}$ $\overline{15}$				4	$\overline{3}$ $\overline{10}$ $\overline{30}$		
\8	\ $\overline{3}$ $\overline{10}$ $\overline{30}$.				\8	\1 $\overline{3}$ $\overline{5}$ $\overline{15}$.		
10	1				10	2		

1	$\overline{\overline{3}}$ $\overline{30}$				1	$\overline{\overline{3}}$ $\overline{10}$ $\overline{30}$		
\2	\1 $\overline{3}$ $\overline{15}$				\2	\1 $\overline{2}$ $\overline{10}$		
4	2 $\overline{\overline{3}}$ $\overline{10}$ $\overline{30}$				4	3 $\overline{5}$		
\8	\5 $\overline{2}$ $\overline{10}$.				\8	\6 $\overline{3}$ $\overline{15}$.		
10	7				10	8		

1	$\overline{\overline{3}}$ $\overline{5}$ $\overline{30}$
\2	\1 $\overline{3}$ $\overline{10}$ $\overline{30}$
4	3 $\overline{2}$ $\overline{10}$
\8	\7 $\overline{5}$.
10	9

SALARY DISTRIBUTION FOR THE PERSONNEL OF THE TEMPLE OF ILLAHUN

Borchardt has given the translation of the salary distributions for the priests of the Illahun temple during the Middle Kingdom* (Table 11.2). The calculations are for payments in loaves of bread and jugs of beer for the various personnel of the temple. For distribution there were 70 loaves, 35 jugs of Sḏꜣ beer, and 115 $\overline{2}$ jugs of Ḥpnw beer. The unit of distribution was one forty-second part of each of these quantities, calculated and written by the scribe as 1 $\overline{3}$ loaves, $\overline{3}$ $\overline{6}$ Sḏꜣ beer, and 2 $\overline{3}$ $\overline{10}$ Ḥpnw beer. This shows an error by the clerk in the portions of Ḥpnw beer; it should be 2 $\overline{2}$ $\overline{4}$, so that he records $\overline{60}$ of a jug too much. This was a pity, because it made his calculations more

* L. Borchardt, "Salary Distribution for the Personnel of the Temple of Illahun," *Zeitschrift für Ägyptische Sprache*, Vol. 40 (Leipzig, 1902–1903), pp. 113–117.

TABLE 11.2
Salary distributions of the personnel of the Temple of Illahun, based upon Borchardt's translation (*Z. Ägypt. Spr.*, Vol. 40, pp. 113–117).

Personnel	Number of Portions 42	Loaves of Bread 1 $\overline{\overline{3}}$	Jugs of Sḏ³ Beer $\overline{\overline{3}}$ $\overline{6}$	Jugs of Ḥpnw Beer 2 $\overline{\overline{3}}$ $\overline{10}$	Corrected Ḥpnw Beer 2 $\overline{2}$ $\overline{4}$
The temple director	10	16 $\overline{\overline{3}}$	8 $\overline{\overline{3}}$	27 $\overline{\overline{3}}$	27 $\overline{2}$
Head lay priest	3	5	2 $\overline{2}$	8 $\overline{5}$ $\overline{10}$	8 $\overline{4}$
Head reader	6	10	5	16 $\overline{2}$ $\overline{10}$	16 $\overline{2}$
Scribe of the temple	1 $\overline{3}$	2 $\overline{6}$ $\overline{18}$	1 $\overline{9}$	3 $\overline{\overline{3}}$ $\overline{45}$	3 $\overline{\overline{3}}$
Usual reader	4	6 $\overline{\overline{3}}$	3 $\overline{\overline{3}}$	11 $\overline{15}$	11
Wtw priest	2	3 $\overline{\overline{3}}$	1 $\overline{\overline{3}}$	5 $\overline{2}$ $\overline{30}$	5 $\overline{2}$
Imi ist c priest	2	3 $\overline{\overline{3}}$	1 $\overline{\overline{3}}$	5 $\overline{2}$ $\overline{30}$	5 $\overline{2}$
Ibh priests (3)	6	10	5	16 $\overline{2}$ $\overline{10}$	16 $\overline{2}$
Royal priests (2)	4	6 $\overline{\overline{3}}$	3 $\overline{\overline{3}}$	11 $\overline{15}$	11
Md ꜣ w	1	1 $\overline{\overline{3}}$	$\overline{\overline{3}}$ $\overline{6}$	2 $\overline{\overline{3}}$ $\overline{10}$	2 $\overline{2}$ $\overline{4}$
Thur guardians (4)	1 $\overline{3}$	2 $\overline{6}$ $\overline{18}$	1 $\overline{9}$	3 $\overline{\overline{3}}$ $\overline{45}$	3 $\overline{\overline{3}}$
Night watchmen (2)	$\overline{\overline{3}}$	1 $\overline{9}$	$\overline{2}$ $\overline{18}$	1 $\overline{2}$ $\overline{\overline{3}}$ $\overline{90}$	1 $\overline{3}$ $\overline{6}$
Temple worker	$\overline{\overline{3}}$	$\overline{2}$ $\overline{18}$	$\overline{4}$ $\overline{36}$	$\overline{\overline{3}}$ $\overline{4}$ $\overline{180}$	$\overline{\overline{3}}$ $\overline{4}$
Another worker*	$\overline{\overline{3}}$	$\overline{2}$ $\overline{18}$	$\overline{4}$ $\overline{36}$	$\overline{\overline{3}}$ $\overline{4}$ $\overline{180}$	$\overline{\overline{3}}$ $\overline{4}$
Totals (clerk)	42§	70§	35§	115 $\overline{2}$#	
Totals, without another worker	41 $\overline{\overline{3}}$	69 $\overline{\overline{3}}$ $\overline{9}$	34 $\overline{\overline{3}}$ $\overline{18}$	115 $\overline{6}$ $\overline{9}$	114 $\overline{\overline{3}}$ $\overline{4}$
Totals, including another worker	42	70	35	115 $\overline{2}$	115 $\overline{2}$

* Omitted by the clerk. Or perhaps there should have been two temple workers.
§ Clerical errors, but only if there was in fact only one temple worker.
# If there were two workers, as seems most likely, then this should be 116 $\overline{5}$. Obviously the clerk did not add up all the fractions. He knew what they *ought* to total, and so he just wrote the numbers down without checking.

difficult. These he performed without further error—except that, when checking his totals, in column 4 he put the total as 115 $\overline{2}$ jugs of Ḥpnw beer as he knew it *should* be, instead of 116 $\overline{5}$, which it *in fact* adds up to. The difference is $\overline{2}$ $\overline{5}$, which of course resulted from his original error of $\overline{60}$ jug too much multiplied by 42, which is $\overline{2}$ $\overline{5}$. The clerk cheated a little here, and we cannot tell how he made the distribution, as his records claim that he did; for either he had $\overline{2}$ $\overline{18}$ of a loaf, $\overline{6}$ $\overline{9}$ jugs of Sḏ³ beer, and $\overline{6}$ $\overline{18}$ jugs of Ḥpnw beer left over, or—

if the correction I have made (really an omission corrected)* is the right one—he found that he was $\overline{2}$ $\overline{5}$ short of Ḥpnw beer. Thus, because of his initial error, the last temple worker may have received only $\overline{6}$ $\overline{18}$ jugs of the last kind of beer instead of $\overline{\overline{3}}$ $\overline{4}$ $\overline{180}$ jugs.

No explanation is given for the number of portions allotted to each person. One is surprised to note that the *head reader* was paid twice as much as the *head lay priest*, and the *usual reader*, three times as much as the *scribe of the temple*!

This clerical record interests us not only because it sheds light upon the relative importance of the various temple personnel; it also causes us to ask, were such meticulous fractional divisions actually carried out in bread and beer? And if they were, how were they done? How, for instance, would the distributor measure out a $\overline{45}$, a $\overline{90}$, or a $\overline{180}$ part of a jug of beer? We have already seen how loaves of bread were divided equally among certain numbers of men in the RMP problems 1 to 6 (pp. 105 and 120–124). Then we can check how the fractional parts of the loaves could in actual fact be distributed according to the clerk's calculations recorded in Table 10.2:

Fractional portions	Number
$\overline{\overline{3}}$	4
$\overline{3}$	2
$\overline{2}$	2
$\overline{6}$	2
$\overline{9}$	1
$\overline{18}$	4.

This is a total of 15 fractional portions, and we see that these could be exactly cut from 5 loaves, with no surplus. For the sum of the fractions just given is 5 (2 $\overline{\overline{3}}$ + $\overline{\overline{3}}$ + 1 + $\overline{3}$ + $\overline{9}$ + $\overline{6}$ $\overline{18}$). Whether the scribe knew this or not, we do not know.

Figure 11.1 shows how such a cutting up of loaves could have been done. Indeed, it would appear that the whole of the calculations of Table 11.2 would need to have been completed before any attempt at

---

* I have added the line, "Another worker," or temple worker, so that the total of portions is 42 *exactly*, just as it should be.

distribution was made; if the order of payment was as indicated in Table 11.2, the scribe of the temple would get a portion of the fourth loaf, and a portion of the fifth loaf. How he received $\overline{45}$ of a jug of beer I do not know! But we can hazard a guess regarding the Sd̠ʾ beer by supposing a dipstick with 36 division marks, or perhaps equally-spaced marks on the containers themselves. But even if this sounds plausible, it strains our credulity too far to suppose that with the clerk's fractions for the Ḥpnw beer he would so divide the stick or the container into 180 equal parts and make the division in the same way. Of course, had he not made his original error of division, then he would only have needed a dipstick with 12 divisions, to measure out $\overline{3}$, $\overline{2}$, $\overline{4}$, and $\overline{6}$, and this would have been relatively easy.

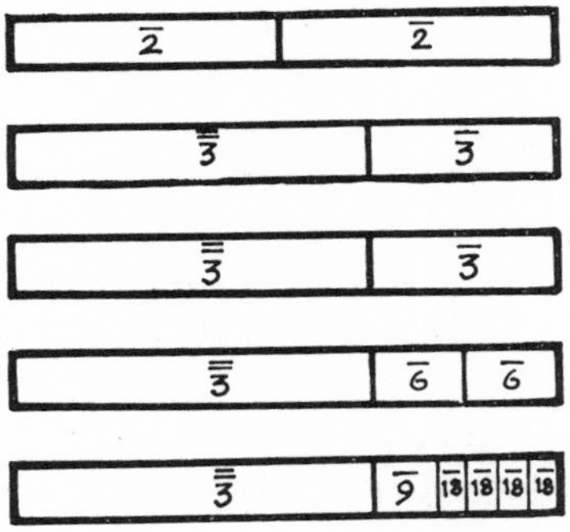

FIGURE 11.1
Cutting up five loaves.

12 PESU PROBLEMS

A *pesu* is a measure of the strengths of beer or bread, *after* either of them is made. It is not a measure of the quality of the barley, wheat, wedyet flour, emmer, besha, spelt-date, dattel, or grain that may have been used to make the beer or bread, although of course all these commodities could vary in quality and strength.* The pesu of the beer or bread made was determined by the Egyptians thus: If one hekat† of grain were used to produce only one loaf or one des-jug of beer,‡ then the pesu of both the bread and the beer was said to be *one*; if one hekat produced two loaves or two des-jugs of beer, then their pesu was said to be *two*; if one hekat produced three loaves or three des-jugs of beer, then their pesu was *three*; and so on, so that the higher the pesu, the weaker the beer or bread, and possibly the smaller the loaf. The relation between the amount of grain used and the pesu of the beer or bread produced was thus:

$$\text{pesu} = \frac{\text{number of loaves or jugs}}{\text{number of hekats of grain}}.$$

Generally speaking, when the Egyptians made beer and bread, they used more grain for their beer than for their bread, or we could say that the same quantity of grain would produce more loaves than des-jugs of beer, which was therefore relatively stronger. In the twenty "pesu-problems" of the RMP and MMP (ten in each), the values of the pesus of the beers lie between 1 and 4, while for the loaves of bread the pesus vary from 5 to as much as 45.

The first of the ten pesu problems of the RMP is number 69:
3 $\bar{2}$ hekats of meal are made into 80 loaves. Find the amount of meal in each loaf and the pesu.

---

* Authorities use various terms to describe pesu; for example, cooking ratio, baking number, or baking value.
† Chace gives 1 hekat = 292.24 cubic inches and Vogel 1 hekat = 4.75 liters; roughly ⅛ bushel.
‡ A des-jug of beer was approximately ⅞ of a pint.

Since there are 320 ro in each hekat, the scribe first found the number of ro in 3 $\bar{2}$ hekats.

\1	320/
\2	640/
\$\bar{2}$	160/
Totals 3 $\bar{2}$	1120 ro.

He then divided 1120 ro by 80. He wrote,

Make thou the operation on 80, for the finding of 1120. The doing as it occurs.

1	80
\10	\800
2	160
\ 4	\320
Totals 14	1120.

The answer is 14 ro in each loaf, or $\overline{32}$ hekat 4 ro.

To find the pesu of each loaf, he only had to divide 80 by 3 $\bar{2}$. He wrote: Make thou the operation on 3 $\bar{2}$ for the finding of 80.

1		3	$\bar{2}$
10		35	
\20		70	/
\ 2		7	/
\ $\bar{3}$		2	$\bar{3}$/
\$\overline{21}$			$\bar{6}$/
\ $\bar{7}$			$\bar{2}$/
Totals 22 $\bar{3}$ $\bar{7}$ $\overline{21}$		80.	

The pesu is 22 $\bar{3}$ $\bar{7}$ $\overline{21}$.

Problems 70 and 71 are like Problem 69, but the remaining seven (Problems 72 to 78) deal with exchange of loaves and beer. Problem 73 is:

100 loaves of pesu 10 are to be exchanged for loaves of pesu 15. How many of these will there be?

The scribe wrote: "Reckon the amount of wedyet flour in these 100 loaves. It is 100 divided by 10, namely, 10 hekats. The number of loaves of pesu 15 from 10 hekats is 15 times 10, namely, 150. This is the number of loaves for the exchange."

The other six problems on the exchange of loaves and beer of different pesus appear at first glance to be similar to Problem 73, but a closer examination reveals some new scribal techniques and arithmetical procedures. We look then at Problems 74 and 76.

PROBLEM 74

1,000 loaves of pesu 5 are to be exchanged, a half for loaves of pesu 10, and a half for loaves of pesu 20. How many of each will there be?

PROBLEM 76

1,000 loaves of pesu 10 are to be exchanged for a number of loaves of pesu 20 and the same number of loaves of pesu 30. How many of each kind will there be?

The scribe's solution for Problem 74 reads as follows: "1,000 loaves of pesu 5 require 200 hekats, and if these are halved, a half of 200 hekats is 100 hekats. Multiply 100 by 10; it makes 1,000, the number of loaves of pesu 10. Multiply 100 by 20; it makes 2,000, the number of loaves of pesu 20. The answer is 1,000 and 2,000 loaves."

The solution for Problem 76 is:

For loaves of pesu 20, the first kind, $\overline{20}$ hekat produces 1 loaf. For loaves of pesu 30, the second kind, $\overline{30}$ hekat produces 1 loaf. Then $\overline{20}\ \overline{30} = \overline{12}$ hekat produces 2 loaves, one of each kind. Then 1 hekat will make 24, or 12 loaves of each kind. The quantity of wedyet flour in the 1,000 loaves of pesu 10 is 100 hekats. Multiply 100 by 12; the result is 1,200, which is the number of loaves of each kind for the exchange.

In order to acquaint ourselves with the scribe's processes in this problem, we set it down in modern terms using what is usually called the *unitary method*.

First, 1,000 loaves of pesu 10 required 100 hekats of flour. Now 1 hekat of flour produces 20 loaves of pesu 20, and 1 hekat of flour produces 30 loaves of pesu 30.

Then 3 hekats of flour produce 60 loaves of pesu 20, and 2 hekats of flour produce 60 loaves of pesu 30.

Therefore 5 hekats of flour produce 60 loaves of each kind, so that 100 hekats of flour produce 60 times 20 loaves of each kind, and the result is 1,200 of each kind of loaf.

A further modern approach now becomes clear to us following the foregoing solution, using the principle of the *harmonic mean*.

The harmonic mean of the two pesus 20 and 30 of the exchange loaves is twice their product divided by their sum, so that the harmonic average of the pesus of the two kinds of loaves considered together is $(2 \times 20 \times 30) \div (20 + 30) = 24$. Now the pesu of the original 1,000 loaves was 10, so that the total number of loaves to be received in exchange is greater in the ratio of 24 is to 10, namely,

$$1,000 \times \frac{24}{10} = 2,400 \text{ loaves.}$$

Then there will be a half of 2,400 or 1,200 loaves of each kind, pesu 20 and pesu 30, received in exchange for 1,000 loaves of pesu 10.

How similar Problems 74 and 76 appear on casual reading, the first asking for *half* for loaves of one pesu and a *half* for loaves of another, while the second problem asks for *equal* numbers of loaves of the two pesus the scribe mentions. But it was a trap for the unwary. It is interesting to note how well Aʿh-mosè chose his two numbers 20 and 30 for the pesus. The choice is reminiscent of the modern problem which asks: If a man drives from one town to another at an average speed of 20 miles per hour, and returns at an average speed of 30 miles per hour, what is his average speed for the double journey? This is like Aʿh-mosè's Problem 76, for the answer to both problems is the harmonic mean of 20 and 30, which is 24 and not 25 (the arithmetic mean). Problem 74 is one in arithmetic means.

The problems of the MMP* dealing with pesus are much the same as those of the RMP, but the scribe was not as careful in his copying

* W. W. Struve. "Mathematischer Papyrus des Museums der Schönen Kunste in Moskau," *Quellen und Studien zur Geschichte der Mathematik*, Part A, Vol. 1 (Berlin, 1930), p. 98.

as was Aʿh-mosè. In addition there are some minor errors of arithmetic, so that the meanings of some of the problems are not entirely
clear.* MMP 21 is one such problem, and the reader is invited to
decide for himself whether or not it resembles Problem 76 of the RMP
and whether the suggestion of a harmonic mean can be found there.

l. 1   Method of calculating the mixing of sacrificial bread.
l. 2   If one names 20 measured as $\overline{8}$ of a hekat and 40 measured as $\overline{16}$
       of a hekat,
l. 3   compute $\overline{8}$ of 20. Result 2 $\overline{2}$.
l. 4   Compute $\overline{16}$ of 40. Result 2 $\overline{2}$.
l. 5   The total of both these halves is 5.
l. 6   Compute the sum of both halves. Result 60.
l. 7   Divide thou 5 by 60.
l. 8   Result $\overline{12}$. Lo! the mixture is $\overline{12}$. You have correctly found it.

In line 7 the scribe divided 5 by 60 where we would have expected 60
divided by 5 giving the pesu of the sacrificial bread as 12. If one hekat
of grain produced 12 loaves of bread then each of these loaves would
have a pesu of 12. But the scribe has expressed this differently by
saying that each loaf contained one twelfth of a hekat of grain, which
is correct. This method of expressing pesu appears to be consistent
with line 2 where the fractions $\overline{8}$ and $\overline{16}$ are written for what we would
call pesus 8 and 16. If then following the scribe's thoughts we think
of fractions only, we come quite naturally to the observation that the
answer $\overline{12}$ is the harmonic mean of the two fractions $\overline{8}$ and $\overline{16}$, being
equal to twice their product divided by their sum.

EXCHANGE OF LOAVES OF DIFFERENT PESUS

The three problems numbered 72, 73, and 75 of the RMP are all
phrased similarly by the scribe, and they treat the same topic.†

RMP 72

100 loaves of pesu 10 are exchanged for loaves of pesu 45. How many
of these are there?

* See Appendix 7.
† Problems 5 and 8 of the MMP are done in the same way.

RMP 73

100 loaves of pesu 10 are exchanged for loaves of pesu 15. How many of these are there?

RMP 75

155 loaves of pesu 20 are exchanged for loaves of pesu 30. How many of these are there?

Since the greater the pesu, the greater the number of loaves from the same quantity of meal, these three problems are very easily solved by simple proportion as follows:

RMP 72

$$\text{No. of loaves} = 100 \times {}^{45}\!/_{10}$$
$$= 10 \times 45$$
$$= 450.$$

RMP 73

$$\text{No. of loaves} = 100 \times {}^{15}\!/_{10}$$
$$= 10 \times 15$$
$$= 150.$$

RMP 75

$$\text{No. of loaves} = 155 \times {}^{30}\!/_{20}$$
$$= 7\ \bar{2}\ \bar{4} \times 30$$
$$= 232\ \bar{2}.$$

This is exactly how the scribe did solve them, that is, all except Problem 72, which for some reason was done in an entirely different manner. Chace remarks that "He arrives at the result in a round-about way," which is very much of an understatement, and although quite true, it adds nothing to our understanding of the scribe's thought processes in arriving at the correct answer. What was the reasoning behind this round-about solution? Was it perhaps a more advanced technique? Was he attempting to introduce some new concept into mathematical methods?

To attempt to answer these questions, we set down the steps in the argument exactly as the scribe gave them (Chace's translation from Vol. 2), line for line.

RMP 72

1.   100 loaves of pesu 10 exchanged for loaves of pesu 45. How many of these loaves are there?

2.   Find the excess of 45 over 10. It is 35. Divide this 35 by 10. You get 3 $\bar{2}$.

3.   Multiply this 3 $\bar{2}$ by 100. Result 350. Add 100 to this 350. You get 450.

4.   Say then that the exchange is 100 loaves of pesu 10

5.   for 450 loaves of pesu 45.

In order to examine logically the steps of the scribe's reasoning, we restate the preceding solution in modern symbolic terms.

1.   If x loaves of pesu p are exchanged for y loaves of pesu q, find y if x, p, and q are known.

2.   Find the excess of q over p. It is $(q - p)$. Divide this $(q - p)$ by p. You get $\left(\dfrac{q - p}{p}\right)$.

3.   Multiply this $\left(\dfrac{q - p}{p}\right)$ by x. Result $\left(\dfrac{q - p}{p}\right)x$. Add x to this. You get $\left(\dfrac{q - p}{p}\right)x + x$.

4.   Say then that the exchange is x loaves of pesu p

5.   for $\left(\dfrac{q - p}{p}\right)x + x$ loaves of pesu q. Then,

$$y = \left(\frac{q - p}{p}\right)x + x$$

$$= \left(\frac{q}{p} - 1\right)x + x$$

$$= x\frac{q}{p} - x + x$$

$$= x \times \frac{q}{p},$$

which is the scribe's formula or method for RMP 73 and 75.

   Now how did the scribe come to think of all this? The only data which he had a priori was the relation,

$$\text{number of hekats of meal} = \frac{\text{number of loaves}}{\text{pesu}}.$$

Following immediately from this, the scribe can write,

$$\frac{x}{p} = \frac{y}{q},$$

whence, $y = x \times q/p$, which is just what he did for RMP 73 and 75.

But for RMP 72, with the same data, to achieve the same steps in his argument as shown in our symbolic transcription of his solution, we are forced to proceed as follows:

Given

$$\frac{x}{p} = \frac{y}{q},$$

then

$$\frac{y}{x} = \frac{q}{p} \qquad \text{(The modern concept of alternando)}$$

and (line 2)

$$\frac{y - x}{x} = \frac{q - p}{p} \qquad \text{(The modern concept of dividendo)};$$

hence (line 3)

$$y - x = \left(\frac{q - p}{p}\right)x,$$

so that (line 3)

$$y = \left(\frac{q - p}{p}\right)x + x$$

$$= \left(\frac{q}{p} - 1\right)x + x$$

$$= x\frac{q}{p} - x + x,$$

therefore,

$$y = x \times \frac{q}{p},$$

exactly as before.

However one looks at this "round-about" method of solution, it is entirely logical and indeed elegant, whether or not the scribe arrived at it by some algebraic or symbolic thought processes, or by some other means. Whatever the true story behind it is, we can only be amazed at such an achievement in 1850 B.C., and suggest that here perhaps, as the scribe wrote it, we are looking at the very earliest example of rhetorical algebra to come to the attention of the historian of mathematics. We cannot avoid introducing the concept of alternando and dividendo in some form or other, because of the scribe's direction in line 2, "Find the excess of q over p."

13 AREAS AND VOLUMES

THE AREA OF A RECTANGLE

From the extant papyri, etc., it is clear that the scribes found the areas of rectangles by multiplying length and breadth as we do today. In Problem 49 of the RMP, the area of a rectangle of length 10 khet (1,000 cubits) and breadth 1 khet* (100 cubits) is found to be $1,000 \times 100 = 100,000$ square cubits. The area was given by the scribe as 1,000 *cubit strips*, which are rectangles, usually of land, 1 khet by 1 cubit (Figure 13.1).

Again in Problem 6 of the MMP, calculation of the area of a rectangle is used in a problem on simultaneous equations. The following text accompanied the rectangle shown on the right in Figure 13.1:

l. 1 Method of calculating a rectangle.
l. 2 If it is said to thee, a rectangle of 12 in the area [is] $\bar{2}$ $\bar{4}$ of the length
l. 3 for the breadth. Calculate $\bar{2}$ $\bar{4}$ until you get 1. Result 1 $\bar{3}$.
l. 4 Reckon with these 12, 1 $\bar{3}$ times. Result 16.
l. 5 Calculate thou its angle [square root]. Result 4 for the length.
l. 6 $\bar{2}$ $\bar{4}$ is 3 for the breadth.

FIGURE 13.1
Rectangles from the RMP and the MMP. *Left*, the 1-khet by 10-khet rectangle of Problem 49 of the RMP. Note that the scribe has made an error in copying, showing the breadth as 2 khet. The working accompanying the figure was done for a breadth of 1 khet. *Right*, a rectangle of area 12 and breadth $\bar{2}$ $\bar{4}$ of the length, from Problem 6 of the MMP. The scribe wrote his correct answers for breadth and width on the figure as shown.

---

* The scribe made a careless copying error, putting 2 khet for 1 khet, and he repeated it in his figure. But there are no errors in his calculations with 1 khet.

Rewriting this in modern form, we would have:
If

$$A = lb = 12 \quad \text{and} \quad b = (\bar{2}\ \bar{4})l, \qquad \text{(from l. 2)}$$

to solve for l, first substitute for b: $l \times (\bar{2}\ \bar{4})l = 12$.

Then

$$l \times l = 12 \div \bar{2}\ \bar{4} \qquad \text{(l. 3)}$$
$$= 12 \times 1\ \bar{3} \qquad \text{(l. 3)}$$
$$= 16. \qquad \text{(l. 4)}$$

Therefore

$$l = 4 \text{ for the length,} \qquad \text{(l. 5)}$$

and

$$\bar{2}\ \bar{4} \text{ of 4 is the breadth 3.} \qquad \text{(l. 6)}$$

THE AREA OF A TRIANGLE

For the area of a triangle ancient Egyptians used the equivalent of the formula $A = \frac{1}{2}bh$. In RMP 51 the scribe shows how to find the area of a triangle of land of side* 10 khet and of base† 4 khet. The scribe took the half of 4, then multiplied 10 by 2 obtaining the area as 20 setats of land. Then in MMP 4 the same problem was stated as finding the area of a triangle of height* 10 and base† 4. As stated by the scribe the method was to calculate with a half of 4 and then to reckon with 10 twice, giving an area of 20. No units such as khets or setats were mentioned.

There have been differences of scholarly opinion among philologists

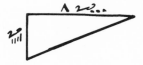 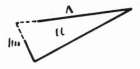

FIGURE 13.2
Triangles accompanying Problem 51 of the RMP (*left*) and Problem 4 of the MMP (*right*).

* The word *meret* (or *meryet*) translated as side or height.
† The word *teper* (or *tepro*) translated as base or mouth.

regarding the precise meaning of *meret* and of *teper*. If *meret* meant the side rather than the height, the area would be in error unless the triangle were right-angled. Scribal sketches of triangles in the papyri suggest that a right-angle may have been intended, but some observers have thought that isosceles triangles with acute vertical angles were meant, and if *meret* was side the area would again be in error by a small amount. However, these differences of opinion are academic; and modern-day historians agree that perpendicular height is meant by the scribe. Thus Peet writes* that he believes "that the Egyptians had found the correct formula, half the base multiplied by the vertical height for the scalene triangle."

D. J. Struik writes,† "The area of a triangle was found as half the product of base and altitude," while Carl B. Boyer has,‡ "Problem 51 (of Ahmes) shows that the area of an isosceles triangle was found by taking half of what we would call the base and multiplying this by the altitude."

THE AREA OF A CIRCLE

In Problem 50 of the RMP, the scribe showed how to find the area of a circle. To explain his method, he assumed a circle of diameter 9 khet as a matter of arithmetical convenience and not because it is a really practical problem. A khet was 100 royal cubits (approximately 57 yards); a *square khet* was called a *setat* (about two-thirds of an acre). The circle the scribe refers to would have an area of over 40 acres and a circumference of nearly a mile.

FIGURE 13.3
Circles drawn by the scribe of the RMP. *Left*, from Problem 41; *right*, from Problem 50.

* T. E. Peet, "Mathematics in Ancient Egypt," *Bulletin of the John Rylands Library*, Vol. 15, No. 2 (Manchester, 1931), p. 430.
† D. J. Struik, *A Concise History of Mathematics*, Dover, New York, 1948, p. 22.
‡ Carl B. Boyer, *A History of Mathematics*, Wiley, New York, 1968, p. 18.

A circle with hieratic inscriptions (Figure 13.3) was included by the scribe in his statement of Problem 50. A translation of Problem 50 is:

Take away $\bar{9}$ of the diameter, namely, 1.
The remainder is 8.
Multiply 8 by 8.
It makes 64.
Therefore it contains 64 setat of land.

Do it thus.

1	9
$\bar{9}$	1.

The remainder is 8.

1	8
2	16
4	32
\8	\64

The area is 64 setat.

The scribe's method of finding the area of a circle can thus be restated: *Subtract from the diameter its one-ninth part, and square the remainder. This is its area.* We ask ourselves how close this is to the true value, and how did the scribe arrive at his formula? If we use the modern value for π, a circle of diameter 9 khet would have an area of 63.6174 setat, so that the Egyptian value is in error by less than 0.6 of one percent.

FIGURE 13.4
Problem 48 of the RMP as the scribe wrote it, including his geometrical illustration.

Problem 48 of the RMP is unique among the 87 problems of the papyrus in that the usual statement of what Aʿh-mosè proposed to do was replaced by a geometrical illustration (Figure 13.4) from which, clearly, the reader is expected to deduce and to understand the nature of the problem. A. B. Chace in his translation of the RMP has written:

PROBLEM 48
Compare the area of a circle and of its circumscribing square.

The circle of diameter 9		The square of side 9	
1	8 setat	\1	9 setat
2	16 setat	2	18 setat
4	32 setat	4	36 setat
\8	64 setat	\8	72 setat
		Total	81 setat

We note that Problem 50 and the first part of Problem 48 appear to be the same, yet their expressions are by no means identical. Indeed, even the numbers as written in the hieratic are not the same. In Problem 48, the 9 is written ⸢⸣, the special sign for 9 khet, also used for 9 hekats of grain, instead of the normal hieratic 9, written ⸢⸣, as in Problem 50 and also in the geometrical diagrams of these problems, where in both it means 9 khet. In Problem 50 the scribe is thinking and writing ordinary numbers simply as units, whereas in Problem 48 he is thinking of setats or square khets as his units, and this is the important circumstance that gives us the clue as to the true purpose of Problem 48. We must conclude that Chace was in error when in translating he interpreted the scribe's working as meaning that the figure is supposed to represent "a circle and its circumscribing square." Chace himself must have had some doubts about the circle, even though he may have thought that it was good enough for a freehand drawing with a scribal reed pen, or fine brush, for he had only to glance at the excellent freehand circle of diameter 9 khet of Problem 50, drawn with the same hand with just two confident sweeps of the pen, to see how good a calligrapher Aʿh-mosè really was. Another circle of diameter 9 cubits in Problem 41 was equally well drawn by Aʿh-mosè (Figure 13.3), and there are other circles in Problems 42 and 43 which support us in the conclusion that the figure within the square of Problem 48 is an octagon with straight sides drawn within a previously drawn square of side 9 khet, and *not* a circle with a square circumscribing it as Chace supposed. Let us look more closely at this octagon of Figure 13.4 now that we agree that it really is one. Was A‘h-mosè intending to inscribe a regular octagon by joining eight points on the four sides of the square? The answer is no! We have no evidence that the ancient Egyptians knew the geometrical construction for determining these points, such that by cutting off the four corners of a square the resulting figure would be an eight-sided figure with all the sides of equal length. And even if A‘h-mosè *did* know of such a construction, he would also know that it would produce a regular octagon of area most certainly greater than that of a circle of diameter 9 units, for it would be an escribed octagon like the square itself. Of course, it would be a much closer approximation to the area of the circle than the square would be, but he would also surely see how he could find a much closer approximation to the circled area, and with a much simpler and more obvious construction. All he would need to do would be to join the adjacent *points of trisection* of the sides. That this is in fact what he did, or was aiming to do, is suggested by his careful choice of a square of side 9 units to allow of easy trisection. Such an octagon would have each pair of opposite sides equal, and in the papyrus except for the top right-hand corner this certainly appears to be the case.

Vogel (1958) came to the same conclusion when he compared our modern formula for the area of a circle, $F = \pi(d/2)^2$, with the equivalent of the Egyptian formula, $F = (8d/9)^2$,[*] from which one derives an Egyptian value for π of $^{256}/_{81}$, which is approximately 3.1605. Vogel then remarks, "Just how this remarkably close approximation was found, we do not know, but we can offer a suggestion on examining the diagram of RMP 48." He then refers to the diagram of Figure 13.5, which he says "... seems to represent a figure whose area approaches the area of a circle inscribed in the square."[†]

[*] K. Vogel, Vorgriechische Mathematik, Vol. 1.
[†] *Ibid.*, "scheint einen Kreis in Annaherung darzustellen, der einem Quadrat einbeschrieben ist."

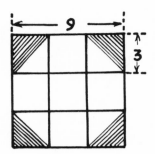

FIGURE 13.5
Vogel's diagram of the inscribed octagon of Problem 48 of the RMP.

Then, he says, the area of the octagon is equal to the original square less the two small squares made up by the four cut-off corners. "Then the area of the octagon is $(81 - 4 \times \frac{9}{2}) = 63$, which would correspond to a square whose side is $\sqrt{63}$, which is approximately $\sqrt{64} = 8$. Thus probably the area of a circle formula $(8d/9)^2$ might have originated."*

ALTERNATE METHOD OF OBTAINING THE FORMULA

By drawing a diagram as that shown in Figure 13.6 on a piece of papyrus, the scribe would conclude that the octagon was pretty closely equal in area to the inscribed circle because some portions of the circle are outside the octagon and some portions of the octagon are outside the circle, and mere observations by the naked eye suggest these are roughly equal. He then would sketch a square of sides representing 9 khet, trisect the sides, join the adjacent points of division, and then by drawing all the lines necessary to actually see or visualize each of the square khets or setats he can count these squares in any way he pleases to find the number of them in the octagon (see Figure 13.6).

Now the Egyptian scribes found the areas of squares and rectangles with ease. Then if the two top shaded corners of 4 $\bar{2}$ setats (or square khets) each, which add to 9 setats, were to replace the top row of 9 setats and if, similarly, the two bottom shaded corners of 9 setats were to replace the left-hand column of 9 setats, then the figure remaining

* *Ibid.*

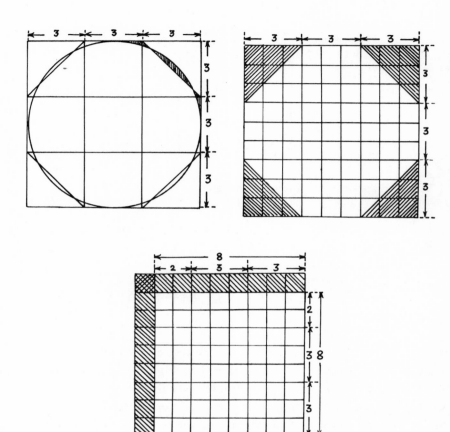

FIGURE 13.6
Diagrams for the alternate method of finding areas of circles. *Left,* the circle and the octagon; *right,* octagonal area marked out in setats or square khets; *below,* the setats from the corners now on two sides.

would be a square instead of an octagon, the area of which the scribe can easily calculate.

The scribe could now properly conclude that the area of the circle inscribed in a square of side 9 khets is very closely equal to the area of a square of side 8 khets.

Therefore, in Problem 48 the scribe finds the total number of setats in this square of eight rows of 8 setats, which he gives as the area of a circle of diameter 9 khets. Of course he certainly knows his method is not *exact*, because he has, so to speak, cut off one setat twice—the one in the top left-hand corner. But his method allows him to find a square nearly equal to a circle, so that we can, "*en caprice*," as it were, credit Aʿh-mosè with being the first authentic circle-squarer in recorded history! Problem 48 shows us that in counting the number of setats in his new square he notes that there are eight rows, each containing 8 setats, so that he multiplies 8 setats by 8, giving him 64 setats. Each step of his calculation gives the product in the special setat hieratic number signs. He does the same in his second multiplication, where he finds the number of setats in the original square to be 81 setats.

Nowhere in this problem does the scribe give the direction, "Take away thou one-ninth of the diameter," as he does in the four other Problems 41, 42, 43, and 50, for this is where he is showing how he discovered his now classical rule. And now we have to meet the critics of ancient Egyptian mathematics, e.g., Sloley,* "Mathematical knowledge in ancient Egypt was essentially practical in character, and must have developed as occasion arose in dealing with problems encountered in daily life." Then Peet, ". . . the Egyptians evolved no better means of stating a formula than that of giving three or four examples of its use, and this is hardly a tribute to the scientific nature of their mathematics."† Such statements as these are legion in histories of mathematics, and they bring to our attention a consideration of the question, What is the nature of proof?

Nowadays one tends to consider that an argument to be rigorous

---

* R. W. Sloley, in *The Legacy of Egypt*, S. R. K. Glanville, ed., Oxford University Press, London, 1942, p. 173.
† T. E. Peet, "Mathematics in Ancient Egypt," *Bulletin of the John Rylands Library*, Vol. 15, No. 2 (Manchester, 1931), p. 439.

must be symbolic. This is not true. A nonsymbolic argument can be quite rigorous, even when given for a particular value of the variable* (in this case it is 9). The only requirements are that the particular value be *typical* and that the generalization to any value be *immediate*. Both of these requirements are satisfied in the argument we have attributed to the scribe in deducing his method, for the value of the diameter, 9, *is* typical, and the generalization to *any* value like 12 or 15 is immediate, as it is also to a number like 20, even though it takes a little longer. The rigor of the argument is implicit in the deduction.

Peet devotes five pages to Egyptian geometry but only eight lines to the circle, even though there are five problems of the RMP dealing with its area. He writes:

The best achievement of the Egyptians in two-dimensional geometry is undoubtedly their close approximation to the area of the circle. They squared eight-ninths of its diameter, giving $256/81r^2$, where r is the radius. Comparing this with our own πr^2 we get for the Egyptian value of π 256/81, or 3 13/81, a very close approximation to 3 1/7, which we find good enough for practical purposes. We have no idea how this result was obtained. The expression of the area as a square suggests a graphic solution.†

Peet's last sentence appears certainly to have been prophetic if the analysis based on Figure 13.6 is accepted.

THE VOLUME OF A CYLINDRICAL GRANARY

The method for finding the volume of a cylinder was, just as it is today, to first find the area of the circular base and then multiply by the height. In Problem 41 of the RMP, for a granary of diameter 9 cubits and height 10 cubits, the scribe subtracted from 9 its $\bar{9}$ part, then multiplied 8 by 8, giving 64, then multiplied by 10, to get 640 cubic cubits. This he expresses in a more convenient unit for grain by multiplying by 1 $\bar{2}$, for there are 1 $\bar{2}$ *khar* in a cubic cubit, so the volume is found to be 960 khar. This last step is analogous to a modern calculation in which the volume would be found in cubic feet, then

---

* I am indebted to Mr. H. Lindgren of Canberra, A.C.T., for suggesting this aspect of Problem 48.
† "Mathematics in Ancient Egypt," p. 434.

afterwards be converted to conventional bushels, or cubic feet to gallons of liquid. This is a first, easy problem, to show the method. Problem 42, which we now discuss, is much more difficult and requires some complicated arithmetic. In this problem we are required to find the contents in khar of a cylindrical granary of diameter 10 cubits and height 10 cubits.

The scribe of the RMP, by the method which he revealed in Problem 41, found the contents to be 1185 $\bar{6}$ $\overline{54}$ khar, and the arithmetic, which involves many operations in unit fractions, will well repay a careful examination. But first we look at a modern solution, using the Egyptian equivalent of π, $^{256}/_{81}$.

$$V = \pi r^2 h$$
$$= {}^{256}/_{81} \times 5^2 \times 10$$
$$= 79^1/_{81} \times 10$$
$$= 790^{10}/_{81} \text{ cubic cubits.}$$

To express this in khar, add to $790^{10}/_{81}$ its half, which is $395^5/_{81}$, which gives a total of $1185^5/_{27}$ khar. Now we express $^5/_{27}$ in unit fractions, in the Egyptian tradition:

$$^5/_{27} = {}^3/_{27} + {}^2/_{27} = \bar{9} \quad \overline{18} \quad \overline{54} = \bar{6} \quad \overline{54} \text{ khar.}$$

Here is a complete statement of the scribe's solution of Problem 42:

Take away $\bar{9}$ of 10 namely 1 $\quad$ $\bar{9}$.
The remainder is 8 $\quad$ $\bar{\bar{3}}$ $\quad$ $\bar{6}$ $\quad$ $\overline{18}$.
Multiply 8 $\quad$ $\bar{\bar{3}}$ $\quad$ $\bar{6}$ $\quad$ $\overline{18}$ by 8 $\quad$ $\bar{\bar{3}}$ $\quad$ $\bar{6}$ $\quad$ $\overline{18}$.
It makes 79 $\quad$ $\overline{108}$ $\quad$ $\overline{324}$ square cubits.

Method of working out:

1. 1	1	8	$\bar{\bar{3}}$	$\bar{6}$	$\overline{18}$	
1. 2	2	17	$\bar{\bar{3}}$	$\bar{9}$		
1. 3	4	35	$\bar{2}$	$\overline{18}$		
1. 4	\ 8	\71	$\bar{9}$			
1. 5	\ $\bar{\bar{3}}$	\ 5	$\bar{\bar{3}}$	$\bar{6}$	$\overline{18}$	$\overline{27}$
1. 6	$\bar{3}$	2	$\bar{\bar{3}}$	$\bar{6}$	$\overline{12}$	$\overline{36}$	$\overline{54}$
1. 7	\ $\bar{6}$	\ 1	$\bar{3}$	$\overline{12}$	$\overline{24}$	$\overline{72}$	$\overline{108}$
1. 8	\ $\overline{18}$	\	$\bar{3}$	$\bar{9}$	$\overline{27}$	$\overline{108}$	$\overline{324}$
1. 9	Total	79	$\overline{108}$	$\overline{324}$.		

l. 10	1	79	$\overline{108}$	$\overline{324}$		
l. 11	10	790	$\overline{18}$	$\overline{27}$	$\overline{54}$	$\overline{81}$
l. 12	$\overline{2}$	395	$\overline{36}$	$\overline{54}$	$\overline{108}$	$\overline{162}$
l. 13	Total	1185	$\overline{6}$	$\overline{54}$	[khar].	

It is left to the reader to examine and explain how the scribe:

found $\overline{9}$ of 10 to be 1 $\overline{9}$;

subtracted 1 $\overline{9}$ from 10 to give 8 $\overline{3}$ $\overline{6}$ $\overline{18}$;

squared the number 8 $\overline{3}$ $\overline{6}$ $\overline{18}$ to get 79 $\overline{108}$ $\overline{324}$;

multiplied this by 10 to get 790 $\overline{18}$ $\overline{27}$ $\overline{54}$ $\overline{81}$; and

multiplied this by 1 $\overline{2}$ to find the contents, 1185 $\overline{6}$ $\overline{54}$ khar.

In Problem 43, the third problem on the contents of a cylindrical granary in the RMP, the scribe proposed to show how the contents may be found directly in khar without having first to calculate the volume in cubic cubits. The scribe did this by means of a very neat and interesting transformation that, correctly interpreted, throws light on the mental processes of the Egyptian scribes. But the copyist Aᶜh-mosè made two errors in his copying of the earlier work from which the RMP was derived. Because of these errors, it was some years before the problem was properly understood upon the work of Schack-Schackenburg, who correctly interpreted a similar problem in the KP.* In that problem, KP IV, 3, the scribe of the KP obtains the contents of a cylindrical granary directly in khar without first finding the volume in cubic cubits, by adding to the diameter its one-third part, squaring this number, then multiplying by two-thirds of the height. In copying a similar calculation in Problem 43 of the RMP, Aᶜh-mosè gave the wrong dimensions for the silo; then he added an extra line of directions, which almost certainly was a carry-over from the two previous problems. The extra line, "take away $\overline{9}$ of the diameter," should have been *replaced* by, "add to the diameter its $\overline{3}$," and *not* been an addition to it. With these corrections made, Problem 43 should read:

A cylindrical granary of diameter 8 and height 6. What is the amount of grain that goes into it?

* F. Griffith, *The Petrie Papyri: Hieratic Papyri from Kahun and Gurob*. University College, London, 1897, pp. 15–18.

Method of working out:

\1			\8	
	$\overline{\overline{3}}$		5	$\overline{3}$
\3	$\overline{3}$		\2	$\overline{3}$
Totals 1	$\overline{3}$		10	$\overline{3}$.

1			10	$\overline{3}$	
\10			\106	$\overline{3}$	
\ $\overline{3}$			\ 7	$\overline{9}$	
Totals 10	$\overline{3}$		113	$\overline{3}$	$\overline{9}$.

1	113	$\overline{\overline{3}}$	$\overline{9}$
2	227	$\overline{2}$	$\overline{18}$
Totals \4	\455	$\overline{9}$	khar, the answer.*

The scribe does not bother to show that 4 is two-thirds of 6 cubits. The second and third lines of doubling show a neat piece of mental reckoning:

1		113	$\overline{\overline{3}}$	$\overline{9}$	
2	226	1	$\overline{3}(\overline{6}$	$\overline{18})$	(Recto 2 ÷ 9)
or,	227		$\overline{2}$	$\overline{18}$	(G rule)
\4	\454		1	$\overline{9}$	
or,	455	$\overline{9}$.			

By taking π equal to $3\frac{1}{7}$, we find that the scribe's answer exceeds the correct value by only 1.8 percent. It is easy enough for us to determine how the scribe's transformation can be arrived at; but it is quite another matter to determine how the ancient Egyptians themselves arrived at the transformation with the much cruder methods and arithmetical tools at their disposal. Despite these handicaps their formula, checked against their standard rule, is found to be correct to within 0.56 percent, also in excess. Let us attempt to see how this may

* We can find modern counterparts to this simplification of a formula in almost any technical manual. For example, in *Shapes and Sections* (a manual of Lysaght's, a subsidiary of B. H. P., the Australian firm, Melbourne, 1937, p. 438), we find the dimensions of a cylindrical tank given in feet, the contents being found directly in imperial gallons rather than in cubic feet, by means of the formula V (in gallons) $= 5\ d^2h$.

have been done, using only those operations that were available to the scribes. The rules are as follows.

STANDARD RULE

Subtract from the diameter its one-ninth.
Multiply this number by itself.
Multiply by the height.
Multiply this number by 1 $\bar{2}$ (or add to it its half).

NEW FORM OF RULE

Add to the diameter its one-third.
Multiply this number by itself.
Multiply by two-thirds of the height.

For our own convenience we put d and h for diameter and height, and we thus have to transform

$$(d - \bar{9}d) \times (d - \bar{9}d) \times h \times (1 \quad \bar{2})$$

into

$$(d + \bar{3}d) \times (d + \bar{3}d) \times \bar{\bar{3}} \times h.$$

Now,

$(d - 9d) \times (d - 9d) \times h \times (1 \quad \bar{2})$

l. 1 $= (\bar{3} + \bar{6} + \overline{18})d \times (\bar{3} + \bar{6} + \overline{18})d \times h \times (1\ \bar{2})$

l. 2 $= (\bar{3} + \bar{6} + \overline{18}) \times 1\ \bar{2} \times (\bar{3} + \bar{6} + \overline{18}) \times (1\ \bar{2})$

                                 $\times\ d \times d \times h \div (1\ \bar{2})$

l. 3 $= (1 + \bar{4} + \overline{12}) \times (1 + \bar{4} + \overline{12}) \times d \times d \times h \times (\bar{3})$

l. 4 $= (1 + \bar{3})d \times (1 + \bar{3})d \times \bar{3}h$

l. 5 $= (d + \bar{3}d) \times (d + \bar{3}d) \times \bar{3}h.$

This is the sequence of operations that could have been made by the scribe (but not, of course, in this form), although it is quite possible that some other sequence might be found to be equally tenable. Line 1 of the sequence, $(1 - \bar{9}) = \bar{3}\ \bar{6}\ \overline{18}$, is shown in line 1 of Problem 42 of the RMP (p. 147). In line 2 of the sequence, multiplication by an extra 1 $\bar{2}$ is balanced by division by 1 $\bar{2}$, which in line 3 becomes a multiplication by $\bar{\bar{3}}$, the well-recognized reciprocal. And again in line 3, multiplying $\bar{3}$ $\bar{6}$ $\overline{18}$ by 1 $\bar{2}$ is

\1		\$\bar{3}$	$\bar{6}$	$\overline{18}$		
\$\bar{2}$		\$\bar{3}$	$\overline{12}$	$\overline{36}$		

$$1 \quad \bar{2} \qquad\qquad (\bar{3} \quad \bar{3})(\bar{6} \quad \overline{12})(\overline{18} \quad \overline{36}),$$

or   1 $\bar{4}$ $\overline{12}$,   (G rule)

and is line 4,   1 $\bar{3}$.   (G rule)

However the scribe may have devised the method of Problem 43, whether it was along the lines I have described or any other, there can be no doubt that, for his era, he was a mathematician of no mean ability.

In Problems 41, 42, and 43 of the RMP, the answers given here in khar are expressed in terms of hekats of grain (where 20 hekats equal 1 khar); but as the numbers are likely to be large for a granary, instead of multiplying by 20 the scribe divides by 20 and calls the answer hundreds of quadruple hekats. This is not of importance regarding the volume of cylinders, but it will come up again when tables of weights and measures are discussed.

THE DETAILS OF KAHUN IV, 3

As in other mathematical workings contained in the Kahun Papyrus, no explanation of what he proposes to do is given by the scribe. All that one reads is the drawing shown in Figure 13.7, accompanied by the following calculation:

\	1	\12	
	$\bar{\bar{3}}$	8	
\	$\bar{3}$	\ 4	
Totals	1 $\bar{3}$	16	
\	1	\ 16	
\10		\160	
\	5	\ 80	
Totals	16	256	

\1	\ 256
2	512
\4	\1024
\3̄	\ 85 3̄
Totals 5 3̄	1365 3̄.

The scribe was here finding the contents of a cylindrical granary of diameter 12 (cubits) and height 8 (cubits). Now if he had followed the standard procedure illustrated by the scribe of the RMP in Problem 50, he would have taken away 9̄ of the diameter 12, squared this, and then multiplied by the height 8. This would have given him the volume 910 6̄ 1̄8̄ in cubic cubits. If he required the answer in khar, he would add a half of 910 6̄ 1̄8̄ to itself, giving 1365 3̄ khar, since there are 1½ khar in a cubic cubit. This is the answer the scribe has written within his freehand-drawn circle to represent the cylindrical granary.

But here the scribe of the KP has used a different technique for the volume of a cylindrical granary, by means of which the contents is determined directly in khar, and it gives evidence of considerable ingenuity on the part of the ancient Egyptian who devised it. In modern terms, it amounts to establishing the "new rule" given on p. 150. The interpretation of the scribe's method puzzled the original translator, F. L. Griffith, who wrote,

It would seem as though the problem had been to find the contents of a circular granary, of which the height and the diameter were 12 and 8 cubits respectively; but if so the method adopted and the result are quite wrong, whether we look for the answer in cubits cubed, in khar, (= 2/3 cubits cubed), or in quadruple heqat.*

FIGURE 13.7
The circle drawn in KP IV, 3, with a translation of the accompanying hieratic.

* *The Petrie Papyri.*

With so little indication given by the scribe, it is not so surprising that the numbers 12 and 8 written outside the circle were erroneously considered by the translator to refer to the height and diameter, rather than the opposite. Thus the scribe shows the working according to the new rule given on p. 150: 1 $\overline{3}$ times 12, giving 16. Then squaring 16 to obtain 256, which the scribe then multiplied by 5 $\overline{3}$, giving the answer, 1365 $\overline{3}$ khar. Had he shown the working, or given a hint that the multiplying factor 5 $\overline{3}$ was $\overline{\overline{3}}$ of the height 8, he would have helped his readers considerably. But finding two-thirds of any numbers at all was mere routine to the Egyptian mathematician, and he no doubt read it off from his two-thirds tables, and thought no more about it. R. C. Archibald, in his bibliography to Chace et al., *The Rhind Mathematical Papyrus*, wrote that Schack-Schackenburg (1899) was the first to explain this problem.

ᛉ ꜹ ᚱᚠ ꜹ ᚱ ꜹ ᚱ

14 EQUATIONS OF THE FIRST AND
SECOND DEGREE

PROBLEMS 24–34 OF THE RMP: SOLUTION OF
FIRST-DEGREE EQUATIONS

As we have seen, many of those who have studied Egyptian mathematical papyri have expressed the opinion that Egyptian mathematics was wholly practical and indeed proceeded by trial-and-error. In order to refute this charge of lack of scientific attitude of mind and the charge that Egyptian mathematicians were concerned only with the practical arithmetic of everyday life, I now exhibit Problems 24–34 of the RMP. These eleven problems deal with the methods of solving equations in one unknown of the first degree. Based upon the order of difficulty and method of solution, these problems fall into three groups.

THE FIRST GROUP

Pr. 24:	A quantity and its $\overline{7}$ added becomes 19. What is the quantity?
Pr. 25:	,,	$\overline{2}$	,,	16.	,,	?
Pr. 26:	,,	$\overline{4}$	,,	15.	,,	?
Pr. 27:	,,	$\overline{5}$	,,	21.	,,	?

Each of these problems is solved by the method known as *false position*, and each deals with abstract numbers unrelated to loaves of bread, hekats of grain, or the areas of fields. It is as if the scribe is showing with four similar problems, but different numbers, a general method of solution for this type of question. The number "falsely assumed" in each case is the simplest that could be chosen, namely, 7, 2, 4, and 5, respectively.

PROBLEM 24

Assume the false answer 7. Then 1 $\overline{7}$ of 7 is 8. Then as many times as 8 must be multiplied to give 19, just so many times must 7 be multiplied to give the correct number. Thus, divide 8 into 19.

	1			8
	\2			\16
	$\bar{2}$			4
	\$\bar{4}$			\ 2
	\$\bar{8}$			\ 1
Totals 2	$\bar{4}$	$\bar{8}$		19.

Now multiply 2 $\bar{4}$ $\bar{8}$ by 7.

	\1		\ 2	$\bar{4}$	$\bar{8}$
	\2		\ 4	$\bar{2}$	$\bar{4}$
	\4		\ 9	$\bar{2}$.	
Totals	7		15	($\bar{2}$ $\bar{2}$) ($\bar{4}$ $\bar{4}$)$\bar{8}$
	7		16	$\bar{2}$ $\bar{8}$.

The answer, then, is 16 $\bar{2}$ $\bar{8}$.

PROBLEM 25

Assume the false answer 2. Then 1 $\bar{2}$ of 2 is 3. Then as many times as 3 must be multiplied to give 16, just so many times must 2 be multiplied to give the correct number. Then divide 3 into 16.

	\1		\ 3
	2		6
	\4		\12
	$\bar{3}$		2
	\$\bar{3}$		\ 1.*
Totals 5	$\bar{3}$		16.

Now multiply 5 $\bar{3}$ by 2.

	1		5	$\bar{3}$
	\2		\10	$\bar{3}$.

The answer is 10 $\bar{3}$.

PROBLEM 26

Assume the false answer 4. Then 1 $\bar{4}$ of 4 is 5. Then as many times as 5 must be multiplied to give 15, just so many times must 4 be multiplied to give the correct number. Then divide 5 into 15.

* Note the method for finding one-third of a number. The scribe first found two-thirds of it, and then halved it, even for the simple case of ⅓ of 3!

\1		\ 5	
\2		\10	
Totals	3	15.	

Now multiply 3 by 4.

1		3	
2		6	
\4		\12	
Totals	4	12.	

The answer is 12.

PROBLEM 27

Assume the false answer 5. Then 1 5̄ of 5 is 6. Then as many times as 6 must be multiplied to give 21, just so many times must 5 be multiplied to give the correct number. Then divide 6 into 21.

\1		\ 6	
\2		\12	
\2̄		\ 3	
Totals 3	2̄	21.	

Now multiply 3 2̄ by 5.

\1		\ 3	2̄
2		7	
\4		\14	
Totals	5	17	2̄.

The answer is 17 2̄.

SIMILAR PROBLEMS FROM OTHER PAPYRI

The following problem (KP LV, 3) is one of a number of mathematical fragments found at Kahun in 1889. It was restored and translated by F. L. Griffith.*

* F. L. Griffith, ed., *The Petrie Papyri: Hieratic Papyri from Kahun and Gurob* (2 vols.), London, 1897.

A half and a quarter are taken away, and 5 remains.
What number says this?
Make thou the remainder after $\bar{2}$ $\bar{4}$ has been taken from 1.
Result $\bar{4}$.
The remainder is $\bar{4}$, if the number was 1.
Then the remainder is $4 \times \bar{4} = 1$, if the number was $4 \times 1 = 4$.
And the remainder is $5 \times 1 = 5$, if the number was $5 \times 4 = 20$.
Therefore, the number that said it was 20.

Though this is a very simple piece of arithmetic, the scribe's reasoning allows us to consider it an example of false assumption with the number assumed being one.

Problem 19 of the Moscow Mathematical Papyrus is as follows. It is not solved by the method of false assumption.

Method of calculating a heap.
1 $\bar{2}$ times together with 4,
it has become 10.
What is this heap?
Compute thou the excess of these 10 over these 4,
it is 6.
Calculate thou with 1 $\bar{2}$ until you find 1.
Result $\bar{\bar{3}}$.
Reckon thou $\bar{\bar{3}}$ of these 6.
Result 4.
Lo! It is 4. You have correctly found it.*

If we express the scribe's reasoning in modern notation, using x to represent the heap, we have

$$1\ \bar{2}x + 4 = 10$$
$$1\ \bar{2}x = 10 - 4$$
$$= 6;$$
$$x = 6 \div 1\ \bar{2}$$
$$= 6 \times \bar{\bar{3}}$$
$$= 4.$$

* W. W. Struve, the translator, has "Du hast richtig gefunden," the MMP scribe's equivalent of the RMP, "Do it thus" or "The doing as it occurs," or even perhaps Euclid's "Quod erat faciendum," Q.E.F.

In view of the abundant evidence that the Egyptian scribes knew as a commonplace statement that the reciprocal of 1 $\bar{2}$ was $\bar{3}$, it is quite surprising to find the scribe of the MMP writing, "Calculate thou with 1 $\bar{2}$ until you find 1."

For each of the Problems 24–27, the scribe supplies a proof that the answer he gives is correct. For Problem 24 it was:

Prove that 1 $\bar{7}$ of 16 $\bar{2}$ $\bar{8}$ is 19.

\1		\	16	$\bar{2}$	$\bar{8}$	
$\bar{7}$			2	($\bar{4}$	$\bar{28}$)	$\bar{14}$ $\bar{56}$
					($2 \div 7 = \bar{4}\ \bar{28}$, Recto RMP)	
		=	2	$\bar{4}$	($\bar{14}$	$\bar{28}$ $\bar{56}$)
\$\bar{7}$		\=	2	$\bar{4}$	$\bar{8}$.	(Pr. 12)

| Totals 1 | $\bar{7}$ | | 18 | $\bar{2}$ | $\bar{4}(\bar{8}$ | $\bar{8}$) |
| 1 | $\bar{7}$ | | 19. | | | |

THE SECOND AND THIRD GROUPS

SECOND GROUP

Two problems constitute the second group. They are:

PROBLEM 28

A quantity and its $\bar{3}$ are added together, and from the sum a third of the sum is subtracted, and 10 remains. What is the quantity?

PROBLEM 29

A quantity and its $\bar{3}$ are added together, $\bar{3}$ of this is added, then $\bar{3}$ of this sum is taken, and the result is 10. What is the quantity?

Both of these problems are discussed in Chapter 16, along with other "think of a number" problems. These problems are *not* "aha" (quantity) problems.

THIRD GROUP

The third group consists of Problems 30–34:

PROBLEM 30

If the scribe says, "What is the quantity of which ($\bar{3}$ $\overline{10}$) will make 10," let him hear!

PROBLEM 31

A quantity, its $\bar{\bar{3}}$, its $\bar{2}$, and its $\bar{7}$ added becomes 33. What is the quantity?

PROBLEM 32

A quantity, its $\bar{\bar{3}}$, and its $\bar{4}$ added becomes 2. What is the quantity?

PROBLEM 33

A quantity, its $\bar{\bar{3}}$, its $\bar{2}$, and its $\bar{7}$ added becomes 37. What is the quantity?

PROBLEM 34

A quantity, its $\bar{2}$, and its $\bar{4}$ added becomes 10. What is the quantity?

   The scribe solved these problems by a different method, that of division. Again, we could suppose that these problems might appear in a modern algebra book, were it not for the choice of the numbers 33, 2, and 37, which lead to enormous unit fractions. So awkward is the mechanical work involved that one is likely to lose sight of the method adopted.

The solutions given by the scribe are:

Pr. 31.   14 $\bar{4}$ $\overline{56}$ $\overline{97}$ $\overline{194}$ $\overline{388}$ $\overline{679}$ $\overline{776}$.

(In modern notation, $14\tfrac{28}{97}$)

Pr. 32.   1 $\bar{6}$ $\overline{12}$ $\overline{114}$ $\overline{228}$.   ($1\tfrac{5}{19}$)

Pr. 33.   16 $\overline{56}$ $\overline{679}$ $\overline{776}$.   ($16\tfrac{2}{97}$)

Pr. 34.   5 $\bar{2}$ $\bar{7}$ $\overline{14}$.   ($5\tfrac{5}{7}$)

A glance at the answers—all of which are quite correct—convinces us that these problems could have had no practical applications. They were meant to illustrate one method for the solution of simple equations of this type, and although they did this, the simplicity of the method has been masked by the complexities of the unit fractions that arise in the process and by the unexpected operations to which the scribe was forced to resort. Thus in one part of the proof of Problem 33, the scribe found that he had to add up 16 fractions, of which the last half-dozen are $\overline{1164}$ $\overline{1358}$ $\overline{1552}$ $\overline{4074}$ $\overline{4753}$ $\overline{5432}$. This is a formidable task in anyone's language or notation. And this the scribe has *apparently* done in his head. In this same problem, the scribe needed to find $\overline{3}$ of the unit fraction $\overline{679}$, which he wrote at once as $\overline{1358}\ \overline{4074}$. This is a remarkable example of the rule given later in the RMP (Problem 61B) referred to in Chapter 4. In Problems 30 to 34 of the RMP the scribe was floundering in a maze of huge fractions. He had chosen his numbers badly, but having started, he continued manfully, and let us remember, successfully, to correct solutions. If the author of one of the earliest mathematical textbooks ever written faltered anywhere, then it was here that he did so. Had he chosen the numbers 97, 19, 194, and 14, in that order, his purpose would have been equally well served, and the mechanical work much simplified. To show how the scribe solved these four problems, we select Problem 34 as the least complicated of them.

PROBLEM 34

A quantity, its $\overline{2}$ and its $\overline{4}$ added becomes 10. What is the quantity? Multiply 1 $\overline{2}$ $\overline{4}$ so as to get 10 (or, divide 10 by 1 $\overline{2}$ $\overline{4}$).

\	1			\ 1	$\overline{2}$	$\overline{4}$			
	2			3	$\overline{2}$				
\	4			\ 7					
\	7								

(Since $4 \times 1\ \overline{2}\ \overline{4} = 7$, $7 \times 1\ \overline{2}\ \overline{4} = \overline{4}$)

	$\overline{4}$	$\overline{28}$			$\overline{4}$		($2 \times 7 = 2 \div 7$
	\$\overline{2}$	$\overline{14}$			$\overline{2}$		$= \overline{4}\ \overline{28}$)
				\ 1.			

Totals 5 $\overline{2}$ 7 $\overline{14}$ 9 $\overline{2}$ ($\overline{4}$ $\overline{4}$)

                       9 ($\overline{2}$ $\overline{2}$)

     5   $\overline{2}$ 7 $\overline{14}$ 10.

The answer is 5 $\overline{2}$ 7 $\overline{14}$. The proof follows:

Find 1 $\overline{2}$ $\overline{4}$ of 5 $\overline{2}$ 7 $\overline{14}$.

\1	\5	$\overline{2}$	7	$\overline{14}$	
\$\overline{2}$	\2	$\overline{2}$	$\overline{4}$	$\overline{14}$	$\overline{28}$
\$\overline{4}$	\1	$\overline{4}$	$\overline{8}$	$\overline{28}$	$\overline{56}$.

Totals 1 $\overline{2}$ $\overline{4}$ 9 ($\overline{2}$ $\overline{8}$) 7 $\overline{14}$ $\overline{14}$ $\overline{28}$ $\overline{28}$ $\overline{56}$.

The scribe refers the last 6 fractions to 56, using red auxiliaries, here rendered in boldface:

7	$\overline{14}$	$\overline{14}$	$\overline{28}$	$\overline{28}$	$\overline{56}$	
8	**4**	**4**	**2**	**2**	**1**	total **21**;
				$\overline{4}$	$\overline{8}$	
				14	**7**	total **21**; therefore they are equal.

Then 9 ($\overline{2}$ $\overline{8}$) $\overline{4}$ $\overline{8}$ = 10.

EQUATIONS OF THE SECOND DEGREE

Two problems in the Berlin Papyrus (Figure 14.1), restored and translated by Schack-Schackenburg,* appear to deal clearly with the solution of simultaneous equations, one being of the second degree. The papyrus is mutilated, so that the restorations, although quite reasonable and plausible, perhaps still remain open to some slight reinterpretation. In essence, Schack-Schackenburg concludes that the scribe proposed to solve the following two sets of equations, expressed in modern algebraic notation:

$$x^2 + y^2 = 100$$
$$4x - 3y = 0 \tag{1}$$

$$x^2 + y^2 = 400$$
$$4x - 3y = 0. \tag{2}$$

The translator's rendering of the first problem (Equations 1) is as follows:

If it is said to thee . . . the area of a square of 100 [square cubits] is equal to that of two smaller squares.
The side of one is $\overline{2}$ $\overline{4}$ the side of the other. Let me know the sides of the two unknown squares.
Always take a square of side 1. Then the side of the other is $\overline{2}$ $\overline{4}$.
Multiply this with $\overline{2}$ $\overline{4}$. It gives $\overline{2}$ $\overline{16}$, the area of the small square.
Then together, these two squares have an area of 1 $\overline{2}$ $\overline{16}$.
Take the square root of 1 $\overline{2}$ $\overline{16}$. It is 1 $\overline{4}$.
Take the square root of this 100 cubits. It is 10.
Divide this 10 by this 1 $\overline{4}$. It gives 8, the side of one square.

The remainder is very much damaged, but what does remain leads Schack-Schackenburg to restore as

Take $\overline{2}$ $\overline{4}$ of these 8. It gives 6, the side of the other square.

* H. Schack-Schackenburg, Der Berliner Papyrus 6619, *Zeitschrift für Ägyptische Sprache*, Vol. 38 (1900), pp. 135–140 and Vol. 40 (1902), p. 65f.

FIGURE 14.1
Berlin Papyrus 6619.

In the papyrus, the squaring of $\bar{2}$ $\bar{4}$ is not shown by the scribe, and neither are the extractions of the square roots of 1 $\bar{2}$ $\overline{16}$ and 100. If these were not done mentally, they were no doubt read off or checked from a table of squares (see Chapter 21). Neither is the working shown for $\bar{2}$ $\bar{4}$ of 8 = 6, and we may justly conclude that the scribe was more concerned to show how his method of false position or false assumption was applied to the solution of equations than to teach the arithmetic of multiplication of fractions and the squaring and square roots of simple fractions.

In the second problem (Equations 2), the amount of restoration is considerable, but as the problem is closely allied to the preceding one the difficulties are lessened. The ratio of the sides of the two smaller squares is given as 2 is to 1 $\bar{2}$, and so these are the sides the scribe assumed instead of, as in the first problem, "Always take a square of side 1." Then he gave the sum of the areas of these two squares as 4 + 2 $\bar{4}$ = 6 $\bar{4}$, the square root of which is given as 2 $\bar{2}$ (no working). The square root of 400 being 20, the scribe divided this 20 by 2 $\bar{2}$ giving 8, which on multiplication by 2 and 1 $\bar{2}$ gave 16 and 12 as the sides of the two smaller squares.

<h3 style="text-align:center">KAHUN LV, 4</h3>

I will now provide a suggested restoration of Problem LV, 4, of the Kahun papyrus, of which the translator F. L. Griffith wrote,

I do not know how to complete this problem, nor what is meant by *hayt*. . . . The word for square root is new and interesting.*

* *The Petrie Papyri.*

My analysis suggests this is a problem of simultaneous equations, one of the second degree, of the general form*

$$xy = A \ldots\ldots (i)$$
$$x = ky \ldots\ldots (ii).$$

l. 31   (Title) ....

l. 32   ....

l. 33   ..... of the hinu ....

l. 34   Make thou that 40, 3 times.

l. 35   The result thereof is 120. Make thou,

l. 36   10 of 120. The result thereof is 12.

l. 37   Make thou that $\bar{2}\ \bar{4}$ to find 1.

l. 38   The result thereof is 1 $\bar{3}$ times. Make thou,

l. 39   that 12, 1 $\bar{3}$ times, the result thereof is 16.

l. 40   Make thou a corner (square root) as 4. Make thou

l. 41 $\bar{2}\ \bar{4}$ of 4. The result thereof is 3.

l. 42   The result thereof is *hayt* (?) 10 of 4 cubits: 3.

Note first of all the following grain measures, as used, e.g., in RMP 41, 42:

a.	10 hinus	= 1 hekat.
b.	20 hekats	= 1 khar.
c.	1 $\bar{2}$ khar	= 1 cubic cubit.

Hence,

d.	300 hinus	= 1 cubic cubit.
e.	200 hinus	= 1 khar.
f.	30 hekats	= 1 cubic cubit.

SUGGESTED RESTORATION OF MISSING LINES OF KAHUN LV, 4,
AND MODERNIZATION OF OTHERS

l. 31   (*Rectangular Granary*). The side is $\bar{2}\ \bar{4}$ of the front. 40 baskets each of 90 hinu,†

l. 32   are poured in, to a depth of 1 cubit. *Find the side and the front.*

* After simplifying the various units, $A = 12$ and $k = \bar{2}\ \bar{4}$.

† A 90-hinu basket would hold slightly more than (1 $\bar{8}$), a bushel of grain, a reasonable load for a laborer to carry.

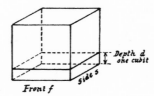

FIGURE 14.2
Rectangular granary of KP LV, 4.

l. 33   Make thou 30 to find 90 "*of the hinu.*" The result is 3.
l. 34   Make thou that 40, 3 times.
l. 35   The result is 120. Make thou,
l. 36 $\overline{10}$ of 120. The result is 12.
l. 37   Divide 1 by $\overline{2}$ $\overline{4}$.
l. 38   The result is 1 $\overline{3}$.
l. 39   Multiply 12 by 1 $\overline{3}$. The result is 16.
l. 40   Find the square root of 16. It is 4, the front.
l. 41 $\overline{2}$ $\overline{4}$ of 4 is 3, the side.
l. 42   The result thereof is (hayt(?) 10 of) *4 cubits: 3 cubits.*
    In modern notation we can write:

The volume of grain is $f \times s \times d = 90 \times 40$ hinu,   (l. 31)

or $f \times s \times 1 = \dfrac{90 \times 40}{30 \times 10}$ cubic cubits   (f,a. l. 32, 33)

$$= 3 \times \frac{40}{10} \qquad \text{(l. 33)}$$

$$= \frac{120}{10}. \qquad \text{(l. 34, 35)}$$

Then, $f \times s = 12, \ldots \ldots (i),$ (l. 36)

and $s = \dfrac{3}{4}f, \ldots \ldots (ii).$ (l. 31)

Substituting for s in (i), $f \times \dfrac{3}{4}f = 12,$ (l. 37)

or $f^2 = 12 \times \dfrac{4}{3},$ (l. 38)

so that $f^2 = 16.$ (l. 39)

Take the square root, $f = 4$ cubits;   (l. 40)

hence, $s = \dfrac{3}{4}f$ ((i) or l. 31)

$$= \frac{3}{4} \times 4; \qquad\qquad (\text{l. } 41)$$

or, $\qquad\qquad s = 3 \text{ cubits}. \qquad\qquad (\text{l. } 41)$

Then the front of the granary is 4 cubits, and the side 3 cubits (l. 42).

As a check we note that

$$90 \times 40 = 3600 \text{ hinu}$$
$$= \frac{3600}{300} \text{ cubic cubits}$$
$$= 12 \text{ cubic cubits}$$
$$= 4 \times 3 \times 1,$$

and since volume $= f \times s \times d$, the depth of grain in the granary is 1 cubit (Figure 14.2).

The approximate modern equivalents of Egyptian dry measure are:

$$1 \text{ hinu} = \frac{4}{5} \text{ pint}.$$
$$1 \text{ hekat} = 1 \text{ gallon} = \frac{1}{8} \text{ bushel}.$$
$$1 \text{ khar} = 2\frac{1}{2} \text{ bushels}.$$
$$1 \text{ cubic cubit} = 3\frac{3}{4} \text{ bushels}.$$

According to Gardiner's dictionary, the *hayt* was the principal multiple of the cubit and was equal to 100 cubits. A partial explanation for "hayt 10 of" (line 42) that I can suggest is that, in the restored line 32, the depth of the grain might have been expressed as $\overline{100}$ of a hayt, which would have been 1 cubit. But this still would not wholly explain the "10 of." There is no suggestion of a hayt in any other part of the problem; and just why the scribe introduced it (if indeed he did) in line 32, I am unable to say.

15 GEOMETRIC AND ARITHMETIC PROGRESSIONS

GEOMETRIC PROGRESSIONS: PROBLEM 79 OF THE
RHIND MATHEMATICAL PAPYRUS

Because of the Egyptians' method of performing all multiplications by continued doubling, it was natural enough that they should be interested in numbers arranged serially, and especially the series 1, 2, 4, 8, 16, . . ., which so often confronted them when they multiplied integers. This series is what we today call a geometrical progression (G.P.).

This particular G.P. whose first term is 1 and whose common multiplier (or common ratio) is 2 has a special property, which the Egyptians must have been aware of and which is today made use of in the

TABLE 15.1
Multipliers written as sums of entries of the geometrical progression 1, 2, 4, 8, 16, ...

Series	Terms to be Added					Resulting Multiplier
1						
2						
	1	2				3
4						
	1		4			5
		2	4			6
	1	2	4			7
8						
	1			8		9
		2		8		10
	1	2		8		11
			4	8		12
	1		4	8		13
		2	4	8		14
	1	2	4	8		15
16						
	1				16	17
		2			16	18
	1	2			16	19
			4		16	20

. . .

design of modern electronic computers. This property is that every
integer can be uniquely expressed as the sum of certain terms of the
series. Thus any integral multiplier, when partitioned in this form,
can be used in Egyptian multiplication. This property is more clearly
seen from Table 15.1.

The importance of this property to the Egyptian scribes lies in the
uniqueness of the partitioning of any multiplier. For example, the
multiplier 26 can be expressed as the sum of terms of this series in *one
way only*, namely, 2 + 8 + 16. It is thus readily understandable that
the Egyptians' attention would quite naturally be directed to the sum
of certain terms of this and other series, and that those properties of
progressions that could be used in subsequent calculations would
interest them deeply. Let us therefore look at one such property of
G.P.'s, to which our attention is drawn by RMP 76. We look at the
series of multipliers that the scribe used in every calculation, viz.,
2, 4, 8, 16, 32, 64, . . . :

The sum of the first 2 terms is $6 = 2 \times (1 + \text{1st. term}) = 6,$

,,　　　　　3　,,　$14 = 2 \times (1 + \text{1st. 2 terms}) = 14,$

,,　　　　　4　,,　$30 = 2 \times (1 + \text{1st. 3 terms}) = 30,$

,,　　　　　5　,,　$62 = 2 \times (1 + \text{1st. 4 terms}) = 62,$

,,　　　　　6　,,　$126 = 2 \times (1 + \text{1st. 5 terms}) = 126,$

. . . .

Then by inductive reasoning the scribe could have concluded that,
in any G.P. whose common multiplier (or common ratio) is the same
as the first term (in this case 2), the sum of any number of its terms is
equal to the common ratio times one more than the sum of the preceding terms. Of course this generalization, on the evidence of only one
such series, would not be sufficient for any scribe, and so it would be
natural to test if the property were true of other series in which, say,
the first term is 3 and the common ratio 3, or the first term 4 and the
common ratio 4, and so on. Then for the G.P. 3, 9, 27, 81, 243, 729, . . . :

The sum of the first 2 terms is $12 = 3 \times (1 + \text{1st. term}) = 12,$

,,　　　　　2　,,　$39 = 3 \times (1 + \text{1st. 2 terms}) = 39,$

,,　　　　　4　,,　$120 = 3 \times (1 + \text{1st. 3 terms}) = 120,$

,,　　　　　5　,,　$363 = 3 \times (1 + \text{1st. 4 terms}) = 363,$

,,　　　　　6　,,　$1{,}092 = 3 \times (1 + \text{1st. 5 terms}) = 1{,}092,$

. . . .

Thus he would find that the property holds for 3 as well as for 2, and very likely this would have been sufficient grounds for an Egyptian mathematician to conclude that it was true for 4, 5, and all other integers, perhaps even fractions.

Let us assume, however, that he wished to further test it for a G.P. whose first term is 7 and whose common ratio is 7. The number 7 often presents itself in Egyptian multiplication because, by regular doubling, the first three multipliers are *always* 1, 2, 4, which add to 7.

Then we will suppose that he tried the G.P. 7, 49, 343, 2,401, 16,807, . . . :

$$\text{The sum of the first 2 terms is } 56 = 7 \times (1 + \text{1st. term})$$
$$= 7 \times 8 = 56,$$
$$\text{,,} \qquad 3 \text{ ,,} \qquad 399 = 7 \times (1 + \text{1st. 2 terms})$$
$$= 7 \times 57 = 399,$$
$$\text{,,} \qquad 4 \text{ ,,} \quad 2,800 = 7 \times (1 + \text{1st. 3 terms})$$
$$= 7 \times 400 = 2,800,$$
$$\text{,,} \qquad 5 \text{ ,,} \quad 19,607 = 7 \times (1 + \text{1st. 4 terms})$$
$$= 7 \times 2,801 = 19,607, \ldots .$$

Thus the procedure holds also for the number 7. Now we refer to Problem 79 of the RMP, which deals with exactly this situation. The scribe, as was usual, gave very little explanation of what he planned to do, and even the few words that he did give are not exactly clear even to the most competent translators of hieratic and hieroglyphic scripts. Nevertheless, whatever the words do signify, the meaning of the sequence of numbers is perfectly clear to us. Problem 79 is quite

TABLE 15.2
Sums of terms of various G.P.'s; such a table may have been known to the scribes.

G.P. whose first term and common ratio is	2	3	4	5	6	7
Sum of 2 terms is	6	12	20	30	42	56
,, 3 ,,	14	39	84	155	258	399
,, 4 ,,	30	120	340	780	1,554	2,800
,, 5 ,,	62	363	1,364	3,905	9,330	19,607

short. The following is Chace's translation:

COL. 1 A HOUSE INVENTORY

1		2,801
2		5,602
4		11,204
	Total	19,607.

COL. 2

Houses	7
Cats	49
Mice	343
Spelt	2,401
Hekat	16,807
Total	19,607.

Clearly column 1 is equivalent to the last step in the preceding argument, 7 × 2,801, but one would have expected to see some reference to where the number 2,801 originated. Either that or a similar multiplication of 400 by 7 ought to be shown, giving 2,800, especially if the scribe was meaning to explain and introduce this shorter method of obtaining his answer. But I suggest this was not the scribe's intention. There seems little doubt that all this detail was earlier known to the Egyptian scribes, and that in Problem 79 use was being made of past experience. It is quite possible that 2,801 had merely to be read off from a table (such as Table 15.2) prepared long before, as (2,800 + 1). There is a minor scribal error in column 2 of the RMP problem, where the scribe wrote 2,301 for spelt, instead of 2,401, yet there is no mistake in the total. This suggests the scribe of the RMP either repeated a previous error or made one himself. Whichever it was, we conclude that the detail of Problem 79 had already been completed before it was set down in the form in which we read it in the RMP.

There have been some fanciful ideas suggested about this problem. One is that here we have the origin of the Mother Goose rhyme,

> As I was going to St. Ives,
> I met a man with seven wives,
> Each wife had seven cats, . . .

All the available evidence for this is here before us, and one is entitled to draw whatever conclusions one wishes. It is indeed tempting to be able to say to a child, "Here is a nursery rhyme that is nearly 4,000 years old!" But is it really? We shall never truly know.

Returning to more mundane things, we can state Problem 79 of the RMP succinctly as follows:

Find the sum of 5 terms of the G.P. whose first term is 7 and whose common ratio is 7.

ARITHMETIC PROGRESSIONS: PROBLEM 40 OF THE RHIND MATHEMATICAL PAPYRUS

In Problem 40 of the RMP, the scribe speaks of an arithmetic progression (A.P.) of 5 terms, in which the sum of the three largest terms is seven times the sum of the two smallest terms. In modern algebra it is usual, though not essential, to think of an A.P. as starting with the smallest term and ending with the largest. The Egyptians reversed this order. For the present discussion we will adopt the modern convention for convenience, and we look first at the very simplest A.P. of all, namely,

$$1, \quad 2, \quad 3, \quad 4, \quad 5.$$

We observe that $3 + 4 + 5 = 12$, which is 4 times $(1 + 2)$. If we look for other series that have similar properties, we can find

$$1, \quad 4, \quad 7, \quad 10, \quad 13,$$

in which $7 + 10 + 13 = 30$, which is 6 times $(1 + 4)$, and

$$1, \quad 14, \quad 27, \quad 40, \quad 53,$$

where $27 + 40 + 53 = 120$, which is 8 times $(1 + 14)$. These are easily found by simple mental arithmetic, but if one were to look for series other than the 4 times, 6 times, and 8 times, say 3 times or 7 times, as the scribe has in RMP 40, then it is a little more difficult, though not so much more, requiring easy fractions such as

$$1, \quad 1\ \bar{2}, \quad 2, \quad 2\ \bar{2}, \quad 3,$$

in which, $2 \quad 2\ \bar{2} \quad 3 = 7\ \bar{2}$, which is 3 times $(1 + 1\ \bar{2})$, and the common difference is $\bar{2}$. Then it is reasonable to suppose that after a little trial and error, the scribe would light on the series

$$1, \quad 6\ \bar{2}, \quad 12, \quad 17\ \bar{2}, \quad 23,$$

where 12   17 $\bar{2}$ 23 $= 52\ \bar{2}$, which is 7 times $(1 + 6\ \bar{2})$, and the common difference is 5 $\bar{2}$. Indeed, this is the series that the scribe used in Problem 40 of the RMP, just as in Problem 79 he used a G.P. whose common ratio is 7, where there are again 5 terms, but the first term is 7, not 1. This is how he framed his problem in Problem 40:

100 loaves for 5 men. $\bar{7}$ of the 3 men above, to the 2 men below. What is the difference of the shares? The doing as it occurs. The share difference being 5 $\bar{2}$.*

1	\23	
1	\17	$\bar{2}$
1	\12	
1	\ 6	$\bar{2}$
1	\ 1	
Total	60.	

\1	60
\$\bar{3}$	40
[Total	100]. (because there are 100, not 60 loaves)

Make thou the multiplication, 1 $\bar{\bar{3}}$.

23		38	$\bar{\bar{3}}$	
17	$\bar{2}$	29	$\bar{6}$	
12		20		
6	$\bar{2}$	10	$\bar{\bar{3}}$	$\bar{6}$
1		1	$\bar{\bar{3}}$	
Totals 60		100.		

The scribe has found each man's share, and not stated the new com-

* This is not the share difference being sought, but the common difference of the A.P. the scribe suggested that one start with.

mon difference. But this is very easily found by subtraction: 1 $\bar{3}$ from 10 $\bar{3}$ $\bar{6}$ is 9 $\bar{6}$.

One would have expected a proof here, following the usual scribal procedure, that the 3 men above did receive 7 times the 2 men below; thus,

$$38\ \bar{3} \qquad 29\ \bar{6} \qquad 20 \text{ is } 87\ \bar{2} \text{ (men above)},$$

and

$$10\ \bar{\bar{3}}\ \bar{6} \qquad 1\ \bar{3} \text{ is } 12\ \bar{2} \text{ (men below)}.$$

Then,

\1		\12	$\bar{2}$
\2		\25	
\4		\50	
Totals 7		87	$\bar{2}.$

But this proof is not given in the papyrus.

We restate the problem in less archaic language.

Divide 100 loaves among 5 men, so that the shares of the 3 highest are together 7 times the shares of the 2 lowest. What is the difference of their shares?

There was no need to specify that the shares are in A.P., for that is inherent in the phrase, "difference of their shares." In his solution, the scribe at once assumed a common difference of 5 $\bar{2}$, because he knew that the A.P.

$$23, \quad 17\ \bar{2}, \quad 12, \quad 6\ \bar{2}, \quad 1$$

exactly fulfills the conditions of the problem, except for the 100 loaves. So then he added up the 5 terms giving 60, which, he noted, will become 100 if multiplied by 1 $\bar{3}$. Therefore he said, "multiply each of the five terms by 1 $\bar{3}$," giving

$$38\ \bar{3}, \quad 29\ \bar{6}, \quad 20, \quad 10\ \bar{\bar{3}}\ \bar{6}, \quad 1\ \bar{\bar{3}},$$

which, as a check, he showed to add up to 100. This is his answer.

The problem asked for the difference of the shares, and presumably the scribe supposed that whoever wishes to know will subtract any two adjacent terms and find it to be 9 $\bar{6}$. But he was more interested in how the 100 loaves were divided among the five men, and clearly this is what he really meant to ask in the problem. For otherwise, if it were *only* the common difference he sought, then he could have found this at once by multiplying the original 5 $\bar{2}$ by 1 $\bar{3}$, giving 9 $\bar{6}$, without bothering to multiply out the five terms of the series by 1 $\bar{3}$.

PROBLEM 64 OF THE RHIND MATHEMATICAL PAPYRUS

Problem 64 of the RMP is:

Example of distributing differences. If it is said to thee, divide *10** hekats of barley among 10 men, so that the difference of each man and his neighbor in hekats of barley, $\bar{8}$ this is, what is each man's share?

The average share [or "*regular*" share in equal distribution] is *1* hekat.
Take away 1 from 10. Remainder is 9.
Half of the difference is found, namely, $\overline{16}$ hekat.
Make up to times 9 [i.e., multiply by 9], there becomes, $\bar{2}$ $\overline{16}$ hekat.
Add this on to the share average.
Take away the $\bar{8}$ hekat for each man until you come to the last.
The doing as it occurs.

$$1\ \bar{2}\ \overline{16},\quad 1\ \bar{4}\ \bar{8}\ \overline{16},\quad 1\ \bar{4}\ \overline{16},\quad 1\ \bar{8}\ \overline{16},\quad 1\ \overline{16},\quad \bar{2}\ \bar{4}\ \bar{8}\ \overline{16},$$
$$\bar{2}\ \bar{4}\ \overline{16},\quad \bar{2}\ \bar{8}\ \overline{16},\quad \bar{2}\ \overline{16},\quad \bar{4}\ \bar{8}\ \overline{16},\quad \text{total } 10 \text{ hekats.}$$

And now I restate the problem in less obscure language:

The sum of 10 terms of an A.P. is 10 and the common difference is $\bar{8}$.
What are the terms of this series?
Find the average value of the terms, 10 ÷ 10 = 1.
The number of differences is one less than the number of terms, = 10 − 1 = 9.
Find half of the common difference, = $\bar{8}$ ÷ 2 = $\overline{16}$.
Multiply 9 by $\overline{16}$ and you get $\bar{2}\ \overline{16}$.
Add this to the average value of the terms, = 1 $\bar{2}\ \overline{16}$.
This is the *highest term*.

* The numbers and fractions in italic type are Horus-eye fractions used solely for measuring grain.

Now *subtract* the common difference $\bar{8}$ nine times, until you reach the *lowest term*.

Then the series is

$$1\ \bar{2}\ \overline{16}, \quad 1\ \bar{4}\ \bar{8}\ \overline{16}, \quad 1\ \bar{4}\ \overline{16}, \quad 1\ \bar{8}\ \overline{16}, \quad 1\ \overline{16}, \quad \bar{2}\ \bar{4}\ \bar{8}\ \overline{16},$$
$$\bar{2}\ \bar{4}\ \overline{16}, \quad \bar{2}\ \bar{8}\ \overline{16}, \quad \bar{2}\ \overline{16}, \quad \bar{4}\ \bar{8}\ \overline{16}.$$

The total is 10 hekats or, as we have it, simply 10.

Alternatively and equally simply, the scribe could have directed us to find the lowest term first and then to *add* the common difference, $\bar{8}$, nine times until we found the highest term; thus,

Subtract this from the average value of the terms, $= \bar{4}\ \bar{8}\ \overline{16}$.
This is the lowest term.
Now *add* the common difference, $\bar{8}$, nine times until you reach the *highest term*.

Then the series is

$$\bar{4}\ \bar{8}\ \overline{16}, \quad \bar{2}\ \overline{16}, \quad \bar{2}\ \bar{8}\ \overline{16}, \quad \bar{2}\ \bar{4}\ \overline{16}, \quad \bar{2}\ \bar{4}\ \bar{8}\ \overline{16}, \quad 1\ \overline{16},$$
$$1\ \bar{8}\ \overline{16}, \quad 1\ \bar{4}\ \overline{16}, \quad 1\ \bar{4}\ \bar{8}\ \overline{16}, \quad 1\ \bar{2}\ \overline{16}.$$

The total is 10 hekats or, as we have it, simply 10.

The scribe preferred the first alternative, because in any series of numbers, it was his custom to write the larger numbers first and the smaller ones following, in descending order of magnitude, just as he did for his fractions. We now follow the scribe's directions word for word, but we substitute for the numbers he used those letters commonly used in a modern algebraic treatment of arithmetic progressions, thus,

a = 1st term (and lowest).
l = last term (and highest)
d = the common difference.
n = the number of terms.
S = the sum of n terms.

The scribe directs,

Find the average value of the n terms, $= S/n$.

The number of differences is one less than the number of terms, $=$ $(n-1)$.

Find half of the common difference, $= d/2$.

Multiply this $(n-1)$ by $d/2$ and you get $(n-1) \times d/2$.

Then either,   or,

Add this to the average value,   *Subtract* this from the average value,

$$\frac{S}{n} + (n-1)\,\frac{d}{2} \qquad\qquad \frac{S}{n} - (n-1)\,\frac{d}{2}.$$

This is the *highest term l*.   This is the *lowest term a*.

Then, $\dfrac{S}{n} + (n-1)\,\dfrac{d}{2} = l,$ Then, $\dfrac{S}{n} - (n-1)\,\dfrac{d}{2} = a,$

or, $\dfrac{S}{n} = l - (n-1)\,\dfrac{d}{2}$ or, $\dfrac{S}{n} = a + (n-1)\,\dfrac{d}{2}$

$$= \frac{1}{2}\,[2l - (n-1)d]; \qquad\qquad = \frac{1}{2}\,[2a + (n-1)d];$$

hence, $S = \dfrac{n}{2}\,[2l - (n-1)d].$ hence, $S = \dfrac{n}{2}\,[2a + (n-1)d].$

How naturally these formulas were deduced! Logically and simply from the scribe's directions! All we have done is to adapt a few elementary algebraic transformations, and we find we have not one, but *two* formulas for the sum of n terms of an A.P. The first is perhaps the less familiar. Indeed, of the 23 algebra texts on my own study shelves, only one* includes it; but they all have the second in exactly the same form as here.

The conclusion is inescapable: hidden in a hieratic script of the Middle Kingdom, expressed in words, without the assistance of any form of algebraic notation at all, is the equivalent of a perfectly well-known modern formula for the sum to n terms of an arithmetic progression. In view of this, it is difficult to accept the recent pronouncement of Professor Morris Kline:

The mathematics of the Egyptians (and Babylonians) is the scrawling of children just learning how to write, as opposed to great literature.†

Scrawling indeed!

* W. E. Paterson, *School Algebra*, Clarendon Press, Oxford, 3rd ed. (1916), p. 385.

† *Mathematics, a Cultural Approach*, p. 14.

KAHUN IV, 3

Tucked away in odd corners of the Kahun Papyri are six relatively small items of mathematical import, which were reproduced, translated, and discussed by Griffith. They were judged to be written about 1800 B.C., and thus would be contemporary with both the RMP and the MMP.

Only three of these so-called mathematical fragments have been clearly explained. KP IV, 2 is a portion of the Recto of the RMP: the odd numbers 3 to 21 divided into 2. KP IV, 3 (columns 13 and 14) deals with the volume of a cylindrical granary, and was first satisfactorily explained by Schack-Schackenburg in 1899. KP LV, 3 solves the equation $\bar{2}x - \bar{4}x = 5$. The remaining three items, KP IV, 3 (columns 11, 12), KP XLV, 1, and KP LV, 4 have not yet been penetrated. Of the first-mentioned, Griffith says, "I must confess I do not see the connexion between the two operations," meaning the columns 11, 12 of KP IV, 3, the second of which appears to be some kind of a series. KP XLV, 1 is a column of seven quite large numbers in decreasing order of magnitude, "not," writes Griffith, "in any fixed proportion, yet it seems probable that they formed part of a considerable mathematical calculation." KP LV, 4 is a vague problem, which Griffith could not complete, and which Schack-Schackenburg in 1900 thought dealt with a quadratic equation. A new word for square root occurred in this problem.

It is KP IV, 3 (columns 11 and 12) that I wish to discuss here and to offer an explanation of its mathematical meaning. My rendering and translation of this problem is shown in Figure 15.1.

At first, in 1966,* I thought the numbers were a solution of, or partly a solution of, a problem that we could express in modern nomenclature as:

In a series of 17 numbers in A.P., whose common difference is double the lowest term, the sum of the 12 largest terms is 110. What is this series?

This indeed does explain both columns of KP IV, 3, though perhaps a little laboriously.

* R. J. Gillings, "Mathematical Fragment from the Kahun Papyrus," *Australian Journal of Science*, Vol. 29, No. 5 (November 1966), p. 126.

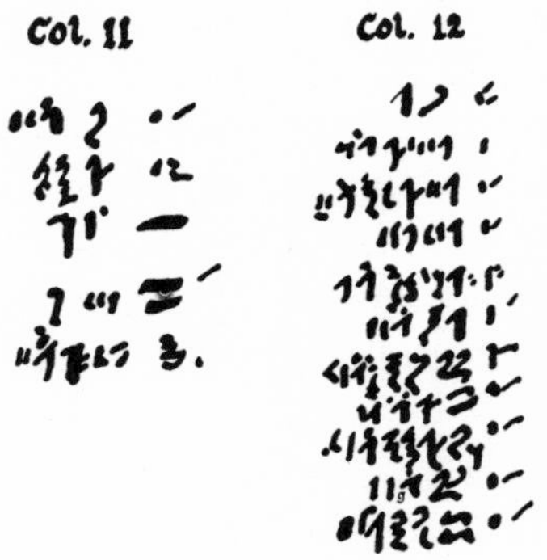

Col. 11

\1	$\bar{3}$ $\overline{12}$
2	$\bar{\bar{3}}$ $\bar{6}$
4	1 $\bar{\bar{3}}$
\8	3 $\bar{\bar{3}}$
Total	3 $\bar{\bar{3}}$ $\overline{12}$

Col. 12

\110			
` 13	$\bar{\bar{3}}$	$\overline{12}$	
\ 12	$\bar{\bar{3}}$	$\bar{6}$	$\overline{12}$
\ 12	$\overline{12}$		
` 11	$\bar{6}$	$\overline{12}$	
\10	$\bar{3}$	$\overline{12}$	
\9	$\bar{3}$	$\bar{6}$	$\overline{12}$
\8	$\bar{\bar{3}}$	$\overline{12}$	
\7	$\bar{\bar{3}}$	$\bar{6}$	$\overline{12}$
\7	$\overline{12}$		
\6	$\bar{6}$	$\overline{12}$	

FIGURE 15.1
Top, KP IV, 3; *bottom,* the translation.

Later, following a closer examination of RMP 64, I developed a modification of this* that appeared to me to be more plausible. It read:

The sum of 12 terms of an A.P. is 110, and the common difference is $\bar{2}$ $\bar{3}$. What is this series?

* R. J. Gillings, "Mathematical Fragment from the Kahun Papyrus, IV, 3," *Australian Mathematics Teacher*, Vol. 23 No. 3 (November 1967), p. 61.

This explanation appeared to receive the approbation of those scholars interested in ancient Egyptian mathematics. But it seems to me that, as we have so few problems on progressions in the extant papyri, those that we do have should be subjected to the very closest scrutiny. So we take a third and closer look at KP IV, 3. Let us first of all clear the numbers of column 12 (the right side of Figure 15.1) of fractions. This we can do by multiplication throughout by 12. Then we obtain the series

   33, 31, 29, 27, 25, 23, 21, 19, 17, 15,

which if continued would give

     13, 11, 9, 7, 5, 3, 1,

where it needs must stop. This is the scribe's familiar series of odd numbers, and we look at it to see if it contains any propositions that would have enabled the scribe to frame problems for solution. The scribe would be at this stage in exactly the same position as a modern textbook writer searching for exercises. He could have observed then that

A	The sum of 4 terms is	120,
B	,,	6 ,,	168,
C	,,	10 ,,	240,
D	,,	12 ,,	264,
E	,,	16 ,,	288.

Of course he could have made similar statements for all the possible sums of the series, but each of the five we have chosen has the factor 12, which property he could have utilized for the framing of problems. Thus if he chose D, he could ask,

If the sum of 12 terms of an A.P. whose common difference is 2 is 264, what is this series?

But if this appeared to him to be perhaps a little elementary, he could have made it somewhat more difficult by dividing the common difference and the sum by the same number, which can be 2, 3, 4, 6, or 12, whichever he chooses, for 12 is an abundant number.* His problem then reads, dividing by 12,

* An *abundant* number is one with many divisors, whose sum exceeds the number itself.

If the sum of 12 terms of an A.P. whose common difference is $\bar{6}$ is 22, what is this series?

But perhaps even this may not be quite challenging enough for his brighter students, so he further multiplies the common difference and the sum by 5, and he has

If the sum of 12 terms of an A.P. whose common difference is $\bar{2}$ $\bar{3}$ is 110, what is this series?

And now he has a problem of the right order of difficulty. Of course, it might well have been that the scribe of KP IV, 3 had himself been set this problem for solution, and that what we read 3,700 years later is his own solution to it; for no explanation of the various steps is written on the papyrus, just as in modern times we find on the papers of most examination candidates. They do not have the time!

I must remark that if the scribe was indeed looking for an interesting problem, he missed a very neat form of it:

If the sum of 10 terms of an A.P. whose common difference is $\bar{2}$ $\bar{3}$ is 100, what is this series?

Indeed, we may justifiably ask, why was this not in fact the original problem being solved? It seems to adequately fit the numbers given in column 12. The only objection to it is column 11 (left half of Figure 15.1), in which a check is made by the scribe. There, the total 3 $\bar{3}$ $\overline{12}$ is the thirteenth term in the series being considered, that is, it is too far forward to be a check on the scribe's work in column 12. This column 11 is our stumbling block. Had the scribe added *all* the products, he would have had

\1	\	$\bar{3}$	$\overline{12}$
\2	\	$\bar{3}$	$\bar{6}$
\4	\1	$\bar{3}$	
\8	\3	$\bar{3}$	
Totals		6	$\bar{4}$.

This finds the tenth term 6 $\bar{4}$, which would have fitted so beautifully with the interpretation we are considering that we would be almost certain that we had here the true answer. But then, how would we explain the 110 at the head of column 12?

We conclude, therefore, that KP IV, 3 is a scribal solution to a problem which we would express in modern terms as

The sum of 12 terms of an arithmetical progression of common difference ($\bar{3}$ $\bar{6}$) is 110. What is this series?

I now apply the directions of Problem 64 of the RMP to KP IV, 3:

Find the average value of the terms, $110 \div 12 = 9$ $\bar{6}$.
One less than the number of terms, $12 - 1 = 11$.
Find half the common difference, $\bar{3}$ $\bar{6} = \bar{3}$ $\overline{12}$.
Make up to 11 times (i.e., multiply by 11), there becomes 4 $\bar{2}$ $\overline{12}$.
Add this to the average value of the terms, 9 $\bar{6}$ 4 $\bar{2}$ $\overline{12} = 13$ $\bar{3}$ $\overline{12}$.
This is the highest term.
Subtract the common difference, ($\bar{3}$ $\bar{6}$), 11 times until you reach the lowest term.
The doing as it occurs.

13 $\bar{3}$ $\overline{12}$, 12 $\bar{3}$ $\bar{6}$ $\overline{12}$, 12 $\overline{12}$, 11 $\bar{6}$ $\overline{12}$, 10 $\bar{3}$ $\overline{12}$, 9 $\bar{3}$ $\bar{6}$ $\overline{12}$,
8 $\bar{3}$ $\overline{12}$, 7 $\bar{3}$ $\bar{6}$ $\overline{12}$, 7 $\overline{12}$, 6 $\bar{6}$ $\overline{12}$,   total 100.

However, the scribe of KP IV, 3 made only 9 of the 11 subtractions. Why did he not complete the subtractions? How can we ever know? But we may surmise that he was checking the progression totals, and when he reached 100, he thought he had finished at 110. Or he may just have got tired of the interminable subtractions. But note that his check multiplication (column 11 in Figure 15.1) totals to the thirteenth term. Including a check mark at 2, a further addition gives 3 $\bar{3}$ $\overline{12} + \bar{3}$ $\bar{6} = 4$ $\bar{2}$ $\overline{12}$, the twelfth term; and a further check mark at 4 gives $8 + 4 + 1 = 3$ $\bar{3}$ 1 $\bar{3}$ $\bar{3}$ $12 = 5$ $\bar{3}$ $\overline{12}$, the eleventh term. These were the two terms he omitted. Had he added all four lines, he would have had $8 + 4 + 2 + 1 = 6$ $\bar{6}$ $\overline{12}$ (the tenth term), which he already had included. All these additions the scribe could quite easily have checked mentally, and no doubt he did this.

I am sure we have found here the true explanation of KP IV, 3, columns 11 and 12, and we find it closely allied to Problem 64 of the RMP, where we have already found hidden the formulas for the sum of n terms of an arithmetic progression (p. 175).

16 "THINK OF A NUMBER" PROBLEMS

Problems 24–27 of the RMP have been discussed in Chapter 14. The two following problems, 28 and 29, were included by Chace, with other *aha* or "quantity" problems. However, I consider these two to be the very earliest examples of *think of a number* problems on record. As long ago as the third century A.D., Diophantus of Alexandria in his *Arithmetica* proposed a class of problems called "find a number" problems, which, however, were mostly concerned with indeterminate equations. In Charles Hutton's* translation of Montucla's (1725–1799) edition of Ozanam (1640–1717), we find the first problem of Chapter 10 to be:

To Tell the Number Thought of by a Person. Desire the person, who has thought of a number, to triple it, and to take the exact half of that triple, if it be even, or the greater half if it be odd. Then desire him to triple that half, and ask him how many times it contains 9; for the number thought of will contain the double of that number of nines, and one more if it be odd.

Vera Sanford† records that Köbel (1514) directed a person to think of a number, add a half of it, add half the sum, divide by nine and tell the result. The result is one fourth of the original number.

For proofs, Hutton and Köbel chose some particular number and then show by arithmetic, not algebra, that the number thought of has been found, and this is exactly what the scribe of the RMP did in his problems 28 and 29.‡

PROBLEM 28 OF THE RHIND MATHEMATICAL PAPYRUS

Chace's translation of Problem 28 (see Figure 16.1) is:

* Charles Hutton, *Recreations in Mathematics and Natural Philosophy*, London, 1840.
† Vera Sanford, *A Short History of Mathematics*, Harrap, London (1930), p. 225.
‡ R. J. Gillings, "Think of a Number: Problems 28, 29 of the RMP," *Mathematics Teacher*, Vol. LIV, No. 2 (Washington, D.C., February, 1961), pp. 97–100.

Two-thirds is to be added. One-third is to be subtracted.
There remains 10.
Make $\overline{10}$ of this, there becomes 1. The remainder is 9.
$\overline{3}$ of it namely 6 is to be added. The total is 15.
$\overline{3}$ of this is 5. Lo! 5 is that which goes out, and the remainder is 10.
The doing as it occurs!

Let us state this in modern terms, adding a few clarifying details:

Think of a number, and add to it its $\overline{3}$. From this sum take away its
$\overline{3}$, and say what your answer is. Suppose the answer was 10.
Then take away $\overline{10}$ of this 10, giving 9. Then this was the number first
thought of.
Proof. If the number were 9, its $\overline{3}$ is 6, which added makes 15. Then $\overline{3}$
of 15 is 5, which on subtraction leaves 10. That is how you do it!

Just as Hutton and Köbel showed how the number originally thought

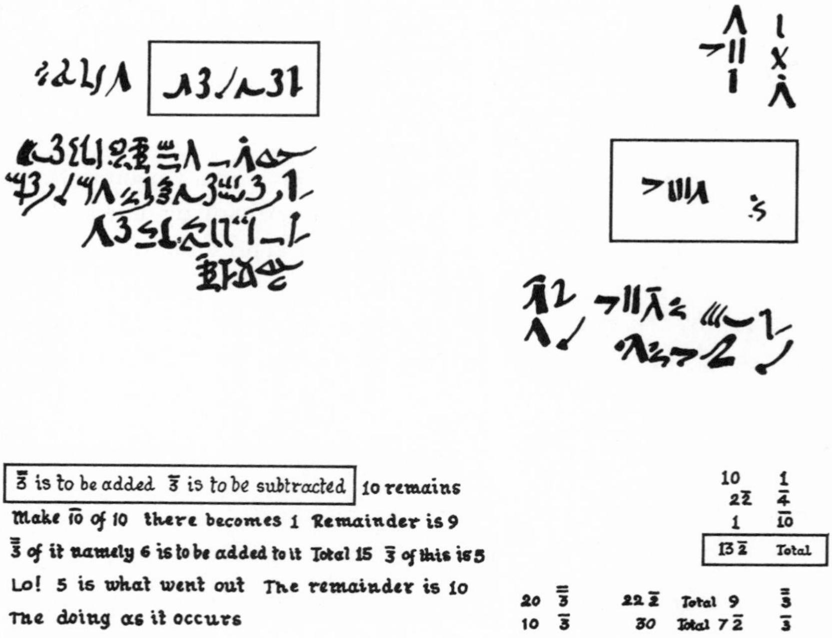

$\overline{3}$ is to be added $\overline{3}$ is to be subtracted	10 remains		
Make $\overline{10}$ of 10 there becomes 1 Remainder is 9			
$\overline{3}$ of it namely 6 is to be added to it Total 15 $\overline{3}$ of this is 5			
Lo! 5 is what went out The remainder is 10			
The doing as it occurs			

FIGURE 16.1
Left, Problem 28 of the RMP, and *right,* Problem 29 of the RMP, as the
scribe wrote them, with translations. Here the lines have been opened up
to accommodate the rectangles, which enclose material drawn in red by
the scribe.

of is found by the exhibition of a specific case, so the scribe of Problem 28 explained how he did it by means of a simple example, in this case using the number 9. Because of the Egyptians' predilection for the $\overline{3}$ fraction and the ease with which they could find $\overline{3}$ of any number, it was only to be expected that these numbers would be used in framing a "think of a number" problem to illustrate the "entrance into the knowledge of all existing things and all obscure secrets."* The scribe could not have chosen a number simpler than 9 to show how his "magic" worked. But to convince his audience, he would have followed with another example. Let us say the number thought of was 54. Then $\overline{3}$ of 54 is 36, which being added is 90. One third of this is 30, which being subtracted leaves 60, which was the answer told to the scribe. All he needed to do now was to subtract from 60 its tenth part, 6, and he at once knew the number thought of was 54.

I must add that neither Peet nor Chace gave this interpretation in their translation and commentaries on the RMP. Peet, for example, suggested that the last line, "the doing as it occurs"† or "do it thus"‡ really belongs to the next problem, and Chace said, "the solution does not seem to be complete." They both regarded it as of the same class as the four preceding problems of the RMP. In no other problem of the RMP do the words "do it thus" occur at the end instead of the beginning of a problem, as both Peet and Chace have pointed out. This I submit gives us the clue to the real intention of the scribe. Having disclosed his "obscure secret," he concluded with—in modern phraseology—"And that is how you do it."

PROBLEM 29 OF THE RHIND MATHEMATICAL PAPYRUS

Problems 28 and 29 stand side by side in the papyrus, and Problem 29 (Figure 16.1) is clearly of exactly the same type as 28. For this reason, the explanatory detail of 29 is the same as that of 28, and so the scribe did not bother to repeat it. If we restore this detail,§ following 28 as closely as possible, we have

* From the scribe's introduction to the RMP.
† Chace's translation in Vol. 2 of *The Rhind Mathematical Papyrus*.
‡ Chace's translation in Vol. 1 of *The Rhind Mathematical Papyrus*.
§ As Peet (1923) and Chace (1927) also did, but not in quite the same words.

Think of a number, and add to it its $\overline{3}$. To this sum add its $\overline{3}$. Find $\overline{3}$ of this result, and say what your answer is. Suppose the answer was 10. Then add $\overline{4}$ and $\overline{10}$ to this 10, giving 13 $\overline{2}$. Then this was the number first thought of.

Proof. If the number were 13 $\overline{2}$, its $\overline{3}$ is 9, which added makes 22 $\overline{2}$. Then $\overline{3}$ of 22 $\overline{2}$ is 7 $\overline{2}$, which added makes 30. Then $\overline{3}$ of this 30 is 10. That is how you do it!

In the papyrus the scribe started with his student's answer which he said was 10, adding to it its $\overline{4}$ and $\overline{10}$.

\ 1	\10	
\ $\overline{4}$	\ 2	$\overline{2}$
\$\overline{10}$	\ 1	

Total,   13 $\overline{2}$. (This is the number thought of.)

Next he proved that he correctly divined it.

\1	\13	$\overline{2}$
\$\overline{3}$	\ 9	

Total,   \22 $\overline{2}$

\$\overline{3}$	\ 7	$\overline{2}$

Total,   30

$\overline{3}$	20
\$\overline{3}$	\10.

(This is the answer he was given.)

The scribe did not put any check marks in either of his calculations for these two problems. We note again* his formality in finding one-third of a number. For $\overline{3}$ of 22 $\overline{2}$, he writes at once, 7 $\overline{2}$, yet for $\overline{3}$ of 30, he first finds $\overline{3}$ of 30 to be 20, then halves the 20 for 10, which is $\overline{3}$ of 30.

This then is the second of the "think of a number" problems of the RMP. No other problems of this type occur in any other mathematical papyrus known to me.

---

* See Chapter 4 on the Egyptian $\overline{3}$ Tables.

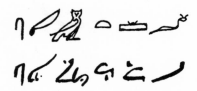

17 PYRAMIDS AND TRUNCATED PYRAMIDS

THE SEKED OF A PYRAMID

We ask ourselves what the extant papyri and ancient records tell us about the ancient Egyptians' knowledge of the geometry and mathematical properties of that most famous geometrical figure of antiquity, the right pyramid. The answer is that they tell us very little. All we know with any certainty is that the scribes knew how to calculate:

1. The seked (slope of the sides) of a pyramid.
2. The volume of a truncated pyramid (or a frustum).
3. The volume of a pyramid.

We know of (3) only because we are certain of (2) from MMP 14. The only other references to calculations on pyramids are Problems 56, 57, 58, 59, and 60 of the RMP. All of these are very simple and deal with the sekeds of various right pyramids. The following is a translation of the text of Problem 56, which is accompanied by the drawing shown in Figure 17.1:

Example of reckoning a pyramid.
Height 250, base 360 cubits.*

What is its seked?

Find $\bar{2}$ of 360, 180.
Divide 180 by 250, $\bar{2}$ $\bar{5}$ $\overline{50}$ cubit.
Now a cubit is 7 palms.
Then multiply 7 by $\bar{2}$ $\bar{5}$ $\overline{50}$.

1			7		
\ $\bar{2}$			\3	$\bar{2}$	
\ $\bar{5}$			\1	$\bar{3}$	$\overline{15}$
\ $\overline{50}$			\	$\overline{10}$	$\overline{25}$
Totals $\bar{2}$	$\bar{5}$	$\overline{50}$		5	$\overline{25}$ palms. This is its seked.

---

* Converting to feet, we find that this problem speaks of a pyramid of height 429 feet and base 618 feet. The actual measurements of the Giza pyramids are: Cheops—481 feet (height), 756 feet (base); Chephren—471 and 708 feet; and Mycerinus—218 and 356 feet.

FIGURE 17.1
The scribe's drawing of a pyramid accompanying Problem 56 of the RMP.

This means that the slope of the triangular faces of this pyramid is
5 $\overline{25}$ palms horizontally for every rise of one cubit in height.

The workers building a pyramid needed to preserve their directions
very carefully in order to obtain the same seked for each subsequent
block of stone, and this may be one reason why the orientation of the
pyramids was so accurately north-south and east-west.

We note that the scribe did not show the division of 180 by 250,
which is simple enough. We note also that in 7 × ($\overline{2}$ $\overline{5}$ $\overline{50}$) no check
marks were made to show which fractions are to be added,* but again
this is no serious omission. What is not so obvious is how he found $\overline{50}$
of 7 to be $\overline{10}$ $\overline{25}$ by finding $\overline{10}$ of 1 $\overline{3}$ $\overline{15}$. No doubt on a papyritic
memo pad, or perhaps even mentally, he had

$$
\begin{array}{ccccc}
\overline{5} & & 1 & \overline{3} & \overline{15} \\
\overline{50} & & \overline{10} & \overline{30} & \overline{150} \\
& & \overline{10} & & \overline{25}. \qquad \text{(G rule†)}
\end{array}
$$

Nor is it obvious how the sum (3 $\overline{2}$ 1 $\overline{3}$ $\overline{15}$ $\overline{10}$ $\overline{25}$) equaled 5 $\overline{25}$
palms. Again, if we could have glanced at his memo pad, we would
no doubt have seen something like this:

$$
\begin{array}{cccccc}
(3 \quad 1) & \overline{2} & \overline{3} & (\overline{10} & \overline{15}) & \overline{25} \\
4 & \overline{2} & (\overline{3} & & \overline{6}) & \overline{25} \qquad \text{(from his tables)} \\
4 & (\overline{2} & & & \overline{2}) & \overline{25} \qquad \text{(from memory or tables)} \\
(4 & & & 1) & & \overline{25} \\
5 & & & & & \overline{25} \text{ palms.}
\end{array}
$$

We may consider the sekeds given in the pyramid problems of the

* We have included them here.
† Or perhaps from $\overline{6}$ $\overline{30}$ = $\overline{5}$ on multiplication by 5.

RMP and MMP to be the cotangents of the angle of slope of the faces of pyramids. Then we may compute these angles and compare them with those actually employed in the Giza pyramids. We obtain:

RMP 56	54° 14′
RMP 57	53° 8′
RMP 58	53° 8′
RMP 59	53° 8′
RMP 60	75° 58′
MMP 14	80° 34′
Cheops	51° 52′
Chephren	52° 20′
Mycerinus	50° 47′.

The remaining problems on pyramids are similar to Problem 56; they read as follows:

RMP 57
The seked of a pyramid is 5 palms 1 finger,* and the base is 140 cubits. What is its height? [93 $\bar{3}$].

RMP 58
The height of a pyramid is 93 $\bar{3}$ cubits, and the base is 140 cubits. What is its seked? [5 palms 1 finger].

RMP 59
The height of a pyramid is 8 cubits, and the base is 12 cubits. What is its seked? [5 palms 1 finger].

RMP 60
A pillar (pyramid?) is 30 cubits high, and its base is 15 cubits. What is its seked? [¼].

THE VOLUME OF A TRUNCATED PYRAMID

The only other problem dealing with pyramids is Problem 14 of the MMP; this problem establishes beyond any doubt that the Egyptians had a standard method for finding the volume of a truncated pyramid. This would represent the very acme of Egyptian mathematical

* Four fingers (or digits) equal one palm.

achievement, except for MMP 10, which some think with justification establishes a formula for finding the area of the curved surface of a hemisphere. The following translation of Problem 14 of the MMP is due primarily to Struve,* though the responsibility for its further translation from German into English is mine (see the scribe's accompanying illustration in Figure 17.2).

Method of calculating a truncated pyramid.
If it is said to thee, a truncated pyramid of 6 *ellen*† in height,

Of 4 *ellen* of the base, by 2 of the top,

Reckon thou with this 4, squaring. Result *16*.
Double thou this 4. Result *8*.
Reckon thou with this 2, squaring. Result *4*.

Add together this 16, with this 8, and with this 4. Result *28*.
Calculate thou $\bar{3}$ of 6. Result *2*.
Calculate thou with 28 twice. Result *56*.

Lo! It is 56! Thou has found rightly.

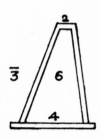
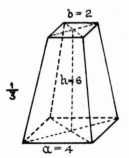

FIGURE 17.2
Left, the scribe's illustration for Problem 14 of the MMP; *right*, a truncated pyramid with symbols replacing particular values.

* W. W. Struve, "Mathematisch Papyrus des Museums in Moskau," *Quellen und Studien zur Geschichte der Mathematik*, Series A, Vol. 1 (Berlin, 1930), p. 135.
† Cubits.

Following the scribe's directions, we may write

$$\begin{aligned}
\text{volume} &= (4 \times 4 + 2 \times 4 + 2 \times 2) \\
&= (16 + 8 + 4) \times \bar{3} \times 6 \\
&= 28 \times 2 \\
&= 56 \text{ cubic cubits.}
\end{aligned}$$

We now replace the particular value of the base 4 by a, the top 2 by b, and the height 6 by h. Then we have

$$\begin{aligned}
V &= (a \times a + b \times a + b \times b) \times \bar{3} \times h \\
&= \frac{h}{3} (a^2 + ab + b^2).
\end{aligned}$$

This is the standard formula for the frustum of a pyramid (see Figure 17.2). But how did the scribes arrive at this?

It has been generally accepted that the Egyptians knew of a method for the volume of a square pyramid, and that it was probably the correct one, $V = \frac{1}{3}ha^2$; but this is nowhere specifically attested, to my knowledge. Problem 14 of the MMP is strong evidence that the Egyptians knew this formula or some equivalent; but it still has not been easy to establish how, even with this powerful tool, they were able to deduce (in a most compact and far from obvious way) the formula for the frustum, which, in the words of Gunn and Peet, "*has not been improved on in 4,000 years.*" This is the question to which we now address ourselves.

It would of course have been a simple operation to construct a hollow pyramid and a hollow rectangular box of the same base and height, to determine that the pyramid had a capacity exactly one-third of the box by simply pouring sand or water. That the Egyptians understood the volume of a rectangular solid to be $l \times b \times d$ is well attested,* so that the volume of an equivalent pyramid would be expressed as one-third of the area of the base times the height, or one-third of the height times the base. In like manner, the two figures could have been made solid of sun-dried Nile River clay and then weighed in the usual Egyptian way. Not so simple is the method of dissection, in which a pyramid is cut up and the parts reformed into a

---

* See, e.g., the Reisner Papyri, and Problems 44, 45, and 46 of the RMP.

rectangular solid whose volume can be easily calculated. None of the dissections I have seen are simple or convincing, but I suggest the following would be within a scribe's capabilities. A right pyramid is constructed of clay or wood, whose perpendicular height is exactly half the side of the square base. This pyramid is then cut into four equal oblique pyramids by two planes passing through the vertex and the midpoints of opposite base lines (Figure 17.3). Then three of these four oblique pyramids fit together to form a cube whose sides are half the base of the pyramid. Therefore in volumes the cube is $\frac{3}{4}$ the pyramid, or the pyramid is $\frac{4}{3}$ the cube. Then the volume of the pyramid is found to be $V = \frac{1}{3}ha^2$.

Another method is to make six congruent *juel** pyramids, that is to say, right pyramids whose heights are half of the sides of the square bases. These would have their triangular sides sloping at 45°, so that an Egyptian would declare their sekeds to be seven palms. These can be put together with their vertices coincident and with their six bases

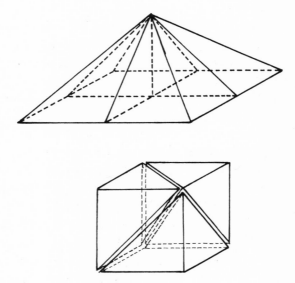

FIGURE 17.3
Dissection of a pyramid whose height is half its base.

* I do not know the origin of this word.

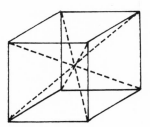

FIGURE 17.4
Six *juel* pyramids fitted together to form a cube.

forming the six faces of a cube (Figure 17.4). An inspection of the dissection immediately yields the correct formula for the volumes of the component pyramids.

Early attempts to ascertain how the Egyptians could have established the equivalent of the formula for finding the volume of a truncated pyramid were made by Gunn and Peet (with Engelbach),* by Kurt Vogel,† by P. Luckey,‡ by W. R. Thomas,§ and more recently by Van der Waerden.‖ In 1964 the dissectionist H. Lindgren of Canberra, Australia, communicated to me a method by which three truncated juel pyramids could be dissected into three rectangular prisms, $(a \times a)$, $(a \times b)$, $(b \times b)$, each of thickness h, which would establish the formula for this particular case. The volume of the truncated pyramid is the volume of the whole pyramid minus the small pyramid cut from the top. Here $k = h + l$, and with the dimensions of MMP 14, $h = l$. Then, from Figure 17.5,

$$\begin{aligned} \text{volume of frustum} &= \tfrac{1}{3}a^2k - \tfrac{1}{3}b^2l \\ &= \tfrac{1}{3}[a^2(h + l) - b^2l] \\ &= \tfrac{1}{3}[a^2h + a^2l - b^2l] \\ &= \tfrac{1}{3}h[a^2 + (a^2 - b^2)]. \end{aligned}$$

In MMP 14, $a = 2b$, so that the scribe would need to cut from the

* *Journal of Egyptian Archaeology*, Vol. 14 (1927).
† *Journal of Egyptian Archaeology*, Vol. 17 (1930).
‡ *Zeitschrift für Mathematik und Naturwissenschaften*, Vol. 41 (1930).
§ *Journal of Egyptian Archaeology*, Vol. 18 (1931).
‖ *Science Awakening* (English translation by Arnold Dresden), Noordhoff, Groningen, 1954, pp. 34, 35.

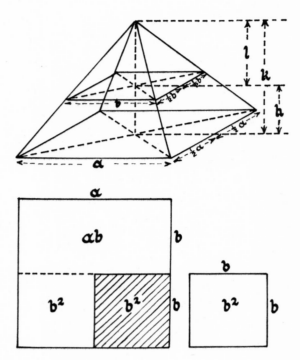

FIGURE 17.5
Volume of a truncated pyramid by dissection. The top edge is half the base.

square of side a a square whose side is b, where $a = 2b$. Then what remains is the rectangle ab and the square b^2, so that $(a^2 - b^2) = (ab + b^2)$. Then,

$$\text{volume of frustum} = \bar{3}h(a^2 + ab + b^2).$$

To check, in the scribal manner, on the validity of this method we take a second case (Figure 17.6), where the base of the pyramid is three times the base of the smaller top pyramid, so that $h = 2l$, and $a = 3b$. Then, from Figure 17.6,

$$
\begin{aligned}
\text{volume of frustum} &= \bar{3}a^2k - \bar{3}b^2l \\
&= \bar{3}[a^2(h + l) - b^2l] \\
&= \bar{3}[a^2h + a^2l - b^2l] \\
&= \bar{3}h[a^2 + a^2\bar{2} - b^2\bar{2}] \\
&= \bar{3}h[a^2 + \bar{2}(a^2 - b^2)].
\end{aligned}
$$

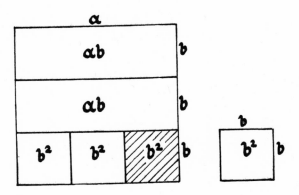

FIGURE 17.6
Second example of the volume of a truncated pyramid by dissection. Here the top edge of the frustum is one-third the base, i.e., the height of the frustum is two-thirds that of the complete pyramid.

To evaluate $\bar{2}(a^2 - b^2)$, which is half the difference of the top and bottom squares of the truncated pyramid, the scribe would have needed to cut from the square of side a a square of side b, where $a = 3b$. There remain the two rectangles, each ab in area, and the two squares, b^2 in area. Thus we have $a^2 - b^2 = 2ab + 2b^2$, and, as before,

$$\text{volume of frustum} = \bar{3}h[a^2 + ab + b^2].$$

In like manner, the scribe could have continued this cutting of smaller pyramids from the top of the larger pyramid to give $a = 3b$, $a = 4b$, $a = 5b$, as far as he wished, or even introduced fractions, so that, for example, $a = 1\ \bar{2}b$, or $b = \bar{3}a$, and thus concluded that his method held in all cases.*

* See Appendix 1.

18 THE AREA OF A SEMICYLINDER AND THE AREA OF A HEMISPHERE

This chapter is based on my article of the same title* in which the views and opinions of Struve, Peet, Neugebauer, and Van der Waerden were considered in attempting to arrive at the conclusion that MMP 10 deals with the area of the curved surface of a hemisphere (and consequently of a sphere) rather than the area of the curved surface of a semicylinder (and consequently of a cylinder).

Struve, the original translator, certainly thought that the scribe's calculations referred to a hemisphere, but Peet thought it more likely a semicylinder was meant; Neugebauer and Van der Waerden inclined toward this latter view. Since my article was published, Van der Waerden has said that he now agrees with Struve and with the arguments I adduced in his favor,† but Neugebauer still feels that the text is not paleographically certain enough to be definitive,‡ so that a further and closer examination of the original papyrus is desirable. Struve died in 1965 and Peet in 1934.

After trying for more than three years I at last received from Moscow a clear photograph of Problem 10 of the MMP from the original papyrus in the Museum of Fine Arts, a copy of which I sent to T. G. H. James, Assistant Keeper of the Department of Egyptian Antiquities of the British Museum, London. A recognized authority on Middle Kingdom hieratic, Mr. James had earlier agreed to lend his services, and in November 1970 he wrote me as follows:

The difficulties reside in the first six lines and are specifically the meaning of *nbt*, the meaning of the words at the end of line 2 and the beginning of line 3, the meaning of *ʿd* and the restoration of the word in line 6. I much prefer to take *nbt* as "basket" and consequently to consider the problem as dealing with the area of a hemisphere, but the difficulties pointed out by Peet for lines 2 and 3 remain. In addition I am not sure that the word *tp-r* means "mouth" or diameter. But I

* In the *Australian Journal of Science*, Vol. 30, No. 4 (October, 1967), pp. 113–116.
† Letter from Van der Waerden, November 11, 1967.
‡ Letter from Neugebauer, December, 8, 1967.

feel in spite of all this that your general interpretation must be along the right lines.

Struve's translation (1930) of Problem 10 of the MMP was published in German;* the following translation into English is mine.

1   Method of calculating a basket.
2   If it is said to thee, a basket with an opening (*mündung*),
3   of 4 $\bar{2}$ in its containing, oh!
4   Let me know its surface.
5   Calculate thou $\bar{9}$ of 9, because the basket
6   is the half of an egg. There results 1.
7   Calculate thou the remainder as 8.
8   Calculate thou $\bar{9}$ of 8.
9   There results $\bar{3}$ $\bar{6}$ $\overline{18}$.
10 Calculate thou the remainder of these 8 left,
11 after taking away these $\bar{3}$ $\bar{6}$ $\overline{18}$. There results 7 $\bar{9}$.
12 Reckon thou with 7 $\bar{9}$, 4 $\bar{2}$ times.
13 There results 32. Lo! This is its area.
14 You have done it correctly.

In 1931, Peet reviewed very carefully and in great detail the whole of Struve's translation and commentary of the Moscow Papyrus.† There he wrote:

If I could believe with Struve that No. 10 involved an approximate determination of the curved area of a hemisphere, this judgment‡ would have to be revised. But I do not! It would be very flattering to the Egyptians, and very important for the history of mathematics, if we could place this brilliant piece of work to their credit.

He then follows in his review with his own rendering of Struve's translation, which is substantially the same as the one I have given, except that his third line reads "of 4 $\bar{2}$ in preservation," instead of "of 4 $\bar{2}$ in its containing." Any real significance in this difference is

---

* W. W. Struve, "The Moscow (Golenishchev) Mathematical Papyrus," *Quellen und Studien zur Geschichte der Mathematik*, Ser. A, Vol. 1 (Berlin, 1930).
† T. E. Peet, A Problem in Egyptian Geometry, *Journal of Egyptian Archaeology*, No. 17 (1931), pp. 100–106.
‡ In 1923, eight years earlier, Peet had examined photographs of the MMP, and he then wrote "It contained nothing, apart from the problem on the truncated pyramid (No. 14), which would greatly modify our conception of Egyptian mathematics."

not immediately apparent. Peet gives his own translations of MMP 10, offering two alternative renderings. And this he finds necessary because of the difficulties of interpreting the scribe's handwriting, in which he says the forms of his signs are "criminally inconsequent"; furthermore, "in some problems he was dealing with a faulty original, or with an original which he did not understand."

Peet's first alternative assumes that the scribe was merely finding the area of a semicircle whose diameter is 9. Now if this were the problem (as in Problem 50 of the RMP), the answer would have been very simply found by subtracting from 9 its one-ninth part, giving 8, squaring this, giving 64, and then halving for a semicircle: 32, as the scribe had. Peet himself thought this interpretation had grave disadvantages but would not dismiss it "out of hand" as a possibility. However, he says it is "unthinkable" that the scribe should find the area of a semicircle by taking $\frac{8}{9}$ of $\frac{8}{9}$ of the diameter (length) and multiplying by the radius (breadth); and so he dismisses the semicircle, in preference for his second alternative of the semicylinder. I have not included Peet's version of MMP 10 assuming a semicircle, as being less plausible than the semicylinder. Peet's second alternative introduces a number that he supposes the scribe inadvertently omitted ($4\ \bar{2}$ of line 2) and makes some further changes from his first alternative, indicated by angular brackets, where he used "circle" for "cylinder."

1   Example of working out a ⟨semicylinder⟩.
2   If they say to you, A ⟨semicylinder of 4 $\bar{2}$* in diameter⟩
3   by 4 $\bar{2}$ in ⟨height⟩, pray,
4   let me know its area. You are to
5   take a ninth of 9, since a ⟨semicylinder⟩
6   is half of a ⟨cylinder⟩; result 1.
7   Take the remainder, namely 8.
8   You are to take a ninth of 8;
9   result $\bar{3}\ \bar{6}\ \overline{18}$. You are to take
10   the remainder of the 8 after ⟨subtracting⟩
11   the $\bar{3}\ \bar{6}\ \overline{18}$; result 7 $\bar{9}$.
12   You are to take 7 $\bar{9}$ 4 $\bar{2}$ times;
13   result 32. See, this is its area.
14   You will find it correct.

* In Peet's first (alternative) translation 4 $\bar{2}$ was replaced by 9.

The replacement of "a basket" by "semicylinder" (lines 1, 2, 3) leads to the inclusion of an extra 4 $\bar{2}$ (line 2), which Peet explains by supposing that in copying the scribe thought that writing 4 $\bar{2}$ twice in succession was unnecessary. It also means that Peet needs to introduce two new terms not previously in evidence: "diameter" (line 2) and "height" (line 3). Then the translation can be paraphrased, "the container is a basket or vessel of cylindrical shape, cut in half, of diameter $4\frac{1}{2}$ and height $4\frac{1}{2}$"* (see Figure 18.1). Now find $\frac{8}{9}$ of $2d$ (lines 5, 6, and 7). Then again, $\frac{8}{9}$ of this $= \frac{8}{9} \times \frac{8}{9} \times 2d$ (lines 8, 9, 10 and 11). Multiply this by h, $= \frac{128}{81}d \times h =$ area (lines 12, 13). Since this may be written

$$A = \frac{1}{2} \times \frac{256}{81}dh,$$

and because $\frac{256}{81}$ is indeed the Egyptian counterpart of the modern π, we have for the area of the curved surface of the semicylinder

$$A = \frac{1}{2}\pi dh.$$

Then if Peet's second alternative is right, we must come inevitably to the conclusion that MMP 10 establishes that the ancient Egyptians knew that the circumference of a semicircle was $\frac{1}{2}\pi d$, and therefore that the circumference of a circle was $C = \pi d$. This would represent a considerable mathematical sophistication for those times and, if

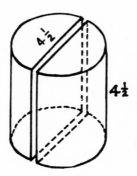

FIGURE 18.1
The cylindrical container whose surface area is calculated in Problem 10 of the MMP, according to Peet.

* We are not told whether the units are cubits or palms.

true, would antedate the Greek Dinostratus by more than 1,400 years. To my knowledge, intimation of this formula is nowhere else attested in the Egyptian mathematical papyri. It is therefore surprising that Peet should write,

The conception of the area of a curved surface does not necessarily argue a very high level of mathematical thought so long as that area is one which, like that of the cylinder, can be directly translated into a plane area by rolling the object along the ground.*

For should this rolling lead to the evaluation of the area of a plane rectangular figure as $2\pi rd$, then it would indeed show a quite high level of mathematical thought. Peet very modestly concludes by saying,

I do not know how many mathematicians I shall convince that this problem deals not with a hemisphere, but with a semicircle or a semi-cylinder.

I now attempt to establish the case for the hemisphere. So far, no one has thought to examine the arithmetical detail of MMP 10, which the scribe probably worked out more fully on his memo pad. Reverting then to Struve's earlier translation (p. 195), we note that the scribe first doubled the diameter 4 $\bar{2}$, getting 9 (line 5), which simplified the next step of finding $\bar{9}$ of 9, which is 1 (lines 5, 6), and then subtracted $(9 - 1) = 8$ (line 7). In lines 8 and 9, he found $\bar{9}$ of 8 to be $\bar{\bar{3}}$ $\bar{6}$ $\overline{18}$, but did not show the working. Following normal technique, he would then have divided 8 by 9. Thus, what the scribe did up to line 7 was to find $\frac{8}{9}$ of twice the diameter, and in lines 8 and 9 he found $\frac{1}{9}$ of this. The division of 8 by 9 would have proceeded as:

1		9		
\ $\bar{\bar{3}}$		\6		
$\bar{3}$		3		
\ $\bar{6}$		\1	$\bar{2}$	
$\bar{9}$		1		
\ $\overline{18}$		\	$\bar{2}$	

Totals $\bar{\bar{3}}$ $\bar{6}$ $\overline{18}$ 8.   Answer $\bar{\bar{3}}$ $\bar{6}$ $\overline{18}$.

* T. E. Peet, A Problem in Egyptian Geometry, *Journal of Egyptian Archaeology*, Vol. 17 (1931), p. 100.

In the next three lines (10, 11, 12), the scribe subtracts this answer from 8, giving him 7 $\bar{9}$, which is his way of finding $\frac{8}{9}$ of 8 to be 7 $\bar{9}$, because he cannot write the fraction $\frac{8}{9}$ in his notation; but he does know that it is the same as $1 - \bar{9}$. The subtraction $8 - (\bar{3}\ \bar{6}\ \overline{18}) = 7\ \bar{9}$ is not shown by the scribe, but this he may well have done mentally or with a few brief jottings, thus:

$$
\begin{array}{llll}
8 = 7 & & \bar{3} & \bar{3} \\
\ \ = 7 & (\bar{6} & \bar{6}) & \bar{3} \\
\ \ = 7 & (\bar{9}\quad \overline{18}) & \bar{6} & \bar{3} \\
\ \ = 7 & \bar{9}\quad (\bar{3} & \bar{6} & \overline{18})
\end{array}
$$

. The remainder is 7 $\bar{9}$.

Finally, in lines 13 and 14 he states that this 7 $\bar{9}$ multiplied by the diameter 4 $\bar{2}$ is 32, the answer to his problem. This multiplication again is not shown, but would have been similar to:

$$
\begin{array}{ll}
\backslash 1 & \backslash 4 \quad \bar{2} \\
\backslash 2 & \backslash 9 \\
\backslash 4 & \backslash 18 \\
\ \ \bar{3} & \ \ 3 \\
\ \ \bar{3} & \ \ 1 \quad \bar{2} \\
\backslash \bar{9} & \backslash \quad\ \bar{2}
\end{array}
$$

Totals 7 $\quad \bar{9}$ $\qquad\qquad$ 32. $\qquad$ Answer 32.

Clearly the scribe was concerned with *method* rather than calculational technique. Let us now condense these arithmetical operations into a simpler form by using modern algebraic notation, using d for the diameter and r for the radius of the hemisphere (see Figure 18.2).

(Line 5). Double 4 $\bar{2}$. Double the diameter $= 2d$.
(Lines 6, 7). Find $\frac{8}{9}$ of this. $\frac{8}{9} \times 2d, = 2 \times \frac{8}{9} \times d$.
(Lines 8, 9, 10, 11). Find $\frac{8}{9}$ of this. $2 \times \frac{8}{9} \times \frac{8}{9} \times d$.
(Line 13). Multiply by d. $2 \times \frac{64}{81} \times d^2 = 2 \times \frac{64}{81} \times (2r)^2$, or

$$A = 2 \times \frac{256}{81}r^2$$
$$A = 2\pi r^2, \qquad \text{where } \pi = \frac{256}{81}.$$

This is indeed the modern formula for the curved surface of a hemisphere. If this interpretation of MMP 10 is the correct one, then the scribe who derived the formula anticipated Archimedes by 1,500 years!

Let us, however, be perfectly clear about both the semicylinder and the hemisphere. In neither case has any *proof* that either $A_{\text{cylinder}} = \frac{1}{2}\pi dh$ or $A_{\text{hemisphere}} = 2\pi r^2$ been established by the Egyptian scribe that is at all comparable with the clarity of the demonstrations of the Greeks Dinostratus and Archimedes. All we can say is that, in the specific case at hand, the mechanical operations performed are consistent with those operations which would be made by someone applying these formulas even though the order and notation might be different. Whether the scribes stumbled upon a lucky close approximation or whether their methods were the results of considered estimations over centuries of practical applications, we cannot of course tell. I find it difficult to accept Peet's suggestion that the area of the curved surface of a semicylinder was meant, but the hemisphere lends itself to plausible speculation. The conventional shapes of Egyptian baskets that are found in murals and other Egyptian art are as shown in Figure 18.3; these were apparently made of rushes or reeds, papyrus, leather, skins, or perhaps even wood. And they must have been made in fairly large numbers, and the art of the basket-maker or weaver must have been one of some consequence in the Egyptian economic world. When one is weaving baskets which are

FIGURE 18.2
The area of a hemisphere for the second interpretation of Problem 10 of the MMP.

FIGURE 18.3
Conventional forms of ancient Egyptian baskets, taken from murals, etc.

roughly hemispherical one requires a quantity of material for the circular plane lid that is about half that required for the basket itself. Since the calculation of the area of a circle was a commonplace operation to the scribes (Problem 50 of the RMP), over a period of years it could have come to be equally commonplace that the curved area of the hemispherical basket was double that of the circular lid.

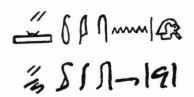

19 FRACTIONS OF A HEKAT

Problem 35 of the RMP reads as follows:

I have gone three times into the hekat-measure, my $\bar{3}$ has been added to me, (and I return), having filled the hekat measure. What is it, that says this?

Call 1 out of 3 $\bar{3}$ [i.e., divide 1 by 3 $\bar{3}$].

	1		3	$\bar{3}$
\	$\overline{10}$	\		$\bar{3}$
\	$\bar{5}$	\		$\bar{3}$
Totals	$\bar{5}$ $\overline{10}$		1.	

PROOF

\	1	\	$\bar{5}$	$\overline{10}$
\	2	\	$\bar{2}$	$\overline{10}$
\	$\bar{3}$	\		$\overline{10}$
Totals	3 $\bar{3}$		1.	

The scribe's answer is that, if a certain container or scoop filled 3 $\bar{3}$ times was required to exactly fill a measure of 1 hekat capacity, then that container must have held $\bar{5}$ $\overline{10}$ hekat. In the RMP, the scribe repeated the above in Horus-eye fractions and ro, where $\bar{5}$ $\overline{10}$ hekat is 96 ro; 96 multiplied by 3 $\bar{3}$ equals 320 ro or 1 hekat. He checked that $\bar{5}$ $\overline{10}$ hekat is 96 ro first.

	1	320
\	$\overline{10}$	32
\	$\bar{5}$	64
Total		96.

In the division of 1 by 3 $\bar{3}$, one step is omitted; we should have

$$
\begin{array}{llll}
& \diagdown\ 1 & \diagdown\ 3 & \bar{3} \\
& \diagdown\ 2 & \diagdown\ 6 & \bar{3} \\
\hline
\text{Totals} & 3 & 10. & \\
\\
\text{Then} & \diagdown\,\bar{10} & \diagdown & \bar{3} \\
& \diagdown\ \bar{5} & \diagdown & \bar{3} \\
\hline
\text{Totals} & \bar{5}\ \bar{10} & 1. &
\end{array}
$$

(If $a \times x = b$
then, $\bar{b} \times x = \bar{a}$)
($\bar{5} = 2 \times \bar{10}$.)

The full proof would be somewhat as follows:

$$
\begin{array}{llllllll}
& \diagdown 1 & 5 & \bar{10} \\
& 2 & (\bar{3} & \bar{15})(\bar{10} & \bar{10}) & & & \text{(Recto } 2 \div 5) \\
& & = \bar{3} & (\bar{10} & \bar{15}) & \bar{10} \\
& & = (\bar{3} & & \bar{6}) & \bar{10} & & \text{(Tables)} \\
& \diagdown 2 = & & \bar{2} & & \bar{10}. & & \text{(Tables)} \\
\hline
\text{again,} & \bar{\bar{3}} & \bar{10} & \bar{30} & \bar{15} & & & \text{(Rule 61B, RMP)} \\
& & = \bar{10} & (\bar{15} & \bar{30}) \\
& & = \bar{10} & & \bar{10} & & & \text{(Tables)} \\
& & = \\
& \diagdown \bar{3} = & & \bar{5} \\
& & & \bar{10}. \\
\hline
\text{Totals } 3 & \bar{3} & \bar{5} & \bar{10} & \bar{2} & \bar{10} & \bar{10} \\
& & \bar{2} & \bar{5} & (\bar{10} & \bar{10}) & \bar{10} \\
& & \bar{2} & (\bar{5} & & \bar{5}) & \bar{10} \\
& & \bar{2} & (\bar{3} & \bar{15}) & & \bar{10} & \text{(Recto } 2 \div 5) \\
& & \bar{2} & (\bar{3} & & \bar{6}) & & (\bar{10}\ \bar{15} = \bar{6}) \\
& & \bar{2} & & & \bar{2} \\
& 3 & \bar{3} & 1. \\
\hline
\end{array}
$$

Many of the detailed steps shown here may well have been done mentally by the scribe or checked on his memo pad.

Problem 36 of the RMP is the same as Problem 35, except that the container-scoop goes 3 $\bar{3}$ $\bar{5}$ times into the hekat-measure. A newer

technique is evident in Problem 36, in that instead of dividing 1 by
3 $\bar{3}$ $\bar{5}$ the scribe divides 30 by 106. Thus,

		1		106	
		$\bar{2}$		53	
	\	$\bar{4}$		\ 26	$\bar{2}$
	\	$\overline{106}$		\ 1	
	\	$\overline{53}$		\ 2	
	\	$\overline{212}$		\	$\bar{2}$
Totals	$\bar{4}$ $\overline{53}$ $\overline{106}$ $\overline{212}$			30.	

The container therefore holds ($\bar{4}$ $\overline{53}$ $\overline{106}$ $\overline{212}$) of a hekat, and the
proof by multiplying by 3 $\bar{3}$ $\bar{5}$ is quite a mammoth calculation,
involving 46 fractional numbers, the largest being $\overline{1060}$. This multiplication is not any more difficult than others, it is only longer.
Repetition with Horus-eye fractions does not occur as in Problem 35.
What most interests us here is the very powerful method the scribes
had at their disposal for what might otherwise be a quite difficult
division with unit fractions. Expressed in algebraic terms, it is
equivalent to

$$a \div b = ax \div bx.$$

The container in Problem 37 of the RMP has to be filled 3 $\bar{2}$ $\overline{18}$ times
to completely fill the one-hekat measure, and therefore 3 $\bar{2}$ $\overline{18}$ has to
be divided into 1. Using the technique of Problem 36, this could be
done easily by dividing 18 by 64:

	1		64
	$\bar{2}$		32
\	$\bar{4}$	\	16
	$\bar{8}$		8
	$\overline{16}$		4
\	$\overline{32}$	\	2
Totals	$\bar{4}$ $\overline{32}$		18. Answer $\bar{4}$ $\overline{32}$ hekat.

But the scribe did not do this. He reverted to his ordinary method and
thus had a long calculation of unit fractions on his hands. His proof
was equally long, but he stuck manfully to his task and then repeated

it with ro, just as he did in Problem 35—the whole entirely without error, which was no mean achievement.

In Problem 38 of the RMP, 3 $\bar{7}$ times the container is required to fill the hekat-measure, so that the scribe had to divide 1 by 3 $\bar{7}$. Had he used his technique of Problem 35, he would have divided 7 by 22 as follows:

1	22		
$\bar{3}$	14	$\bar{3}$	
$\bar{3}$	7	$\bar{3}$	
\ $\bar{6}$	\ 3	$\bar{2}$	$\bar{6}$
\ $\overline{11}$	\ 2		
\ $\overline{22}$	\ 1		
\ $\overline{66}$	\		$\bar{3}$

Totals $\bar{6}$ $\overline{11}$ $\overline{22}$ $\overline{66}$ 7. Answer $\bar{6}$ $\overline{11}$ $\overline{22}$ $\overline{66}$ hekat.

But again he did not use this elegant technique, which clearly he was aware of. But he did something which is closely akin to it, in dividing 1 by 3 $\bar{7}$. He wrote

1	3	$\bar{7}$
$\overline{22}$		$\bar{7}$

in which he was saying that

$$\text{if } a \times x = b$$
$$\text{then } \bar{x} \times b = a,$$

and thus, because he knew that $\bar{7} \times 22 = 3\ \bar{7}$, then

$$\overline{22} \times 3\ \bar{7} = \bar{7}.$$

Then he wrote

	1		3	$\bar{7}$
\	$\overline{22}$	\		$\bar{7}$
\	$\overline{11}$	\	4	$\overline{28}$
\ $\bar{6}$	$\overline{66}$	\	2	$\overline{14}$

Totals $\bar{6}$ $\overline{11}$ $\overline{22}$ $\overline{66}$ 1.

Here the third line comes from doubling the second, so that $2 \times \overline{22} = \overline{11}$, and $2 \times \overline{7} = \overline{4}\ \overline{28}$, from the Recto $2 \div 7$. Again the fourth comes from doubling the third, so that $2 \times \overline{11} = \overline{6}\ \overline{66}$, from the Recto $2 \div 11$, and $2 \times \overline{4}\ \overline{28} = \overline{2}\ \overline{14}$. The addition of the right-hand side column to give 1 is

$$\overline{2} \quad \overline{7} \quad \overline{4} \quad \overline{14} \quad \overline{28} = \overline{2} \quad \overline{4} \quad (\overline{7} \quad \overline{14} \quad \overline{28})$$
$$= \overline{2} \quad \overline{4} \qquad\qquad \overline{4}$$
$$= 1.$$

The value $\overline{7}\ \overline{14}\ \overline{28} = \overline{4}$ is known from tables or from Problem 10 of the RMP and was an equality very well known to the scribes. The proof that follows is also not simple, and when the whole is repeated in Horus-eye fractions and ro, the solution to Problem 38 appears as a rather long and involved calculation, when in fact it could have been a simple one if the scribe had taken care to use his more efficient methods.

CUBIT:

A *cubit* was originally the length of a forearm, from the elbow to the tip of the middle finger. Of course, individuals' limbs varied in length; and two standard cubits came into common use early, the *Royal Cubit* and the *Short Cubit*. The former was the cubit usually meant in measuring in everyday life and was 20.6 inches (more accurately 20.59),* while the short cubit is reckoned to be 17.72 inches, hence the "cubit and an hand breadth":

Behold . . . there was a man . . . with a line of flax in his hand, and a measuring reed . . . of six cubits long by the cubit and an hand breadth; so he measured the breadth of the building, one reed, and the height, one reed.

Then measured he the porch of the gate, eight cubits; and the posts thereof, two cubits.

The foundations of the side chambers were a full reed of six great cubits. . . . And these are the measures of the altar after the cubits: the cubit is a cubit and an hand breadth. And the altar shall be twelve cubits long, twelve broad, square in the four squares thereof. EZEKIEL 40:3, 5, 9; 41:8; 43:13, 16.

In later times the term "cubit" was still used, the Greek cubit being 18.22 inches and the Roman 17.47 inches.† Ezekiel is contemporary with the Babylonian King Nebuchadnezzar and the Egyptian Pharaoh Apries (Hophra of the Bible), who reigned from 589 to 570 B.C.‡

PALM:

The *palm*, or handbreadth, was one-seventh of a cubit, and thus 2.94

* Sir Alan Gardiner, *Egyptian Grammar*, Oxford University Press, London, 3rd edition, 1957, p. 199, has royal cubit = 0.523 meter. See also I. E. S. Edwards, *The Pyramids of Egypt*, Pelican, London, 1952, p. 208; R. W. Sloley in *The Legacy of Egypt*, S. R. K. Glanville, editor, Oxford University Press, London, 1942, p. 176. Both give 1 cubit = 20.62 inches. I adopt Gardiner's value of 20.59 inches.
† Webster's New International Dictionary, 2nd edition, 1934.
‡ Gardiner, *Egypt of the Pharaohs*, Oxford University Press, London, 1961, p. 360; S. R. K. Glanville, editor, *The Legacy of Egypt*, p. 233.

inches if taken from a royal cubit and 2.53 inches from a short cubit. To the nearest tenth of an inch, the royal cubit was 2.9 inches longer than the short cubit. Some authorities state that for the short cubit six instead of seven palms was the equality used. This is plausible, for then the short-cubit palm would be 2.95 inches, very close to the royal-cubit palm of 2.94 inches.

FINGER:

A *finger*, sometimes called a *fingerbreadth* or a *digit*, was one-quarter of a palm or handbreadth. Thus 28 fingers equaled a cubit. It was nearly $3/4$ inch, or 0.735 inch from the royal cubit and slightly less from the short cubit.

HAYT:

The chief multiple of the cubit was the *hayt* (*rod* or *cord*) of 100 cubits.

REMEN:

A *double-remen* was the length of the diagonal of a square whose side was one cubit. Using the royal cubit, which was most commonly the case, a double-remen was therefore 29.1325 inches($\sqrt{2} \times 20.6$), and consequently the *remen* was 14.566 inches. It is thought that the double-remen was used in measuring land, because it enabled areas to be halved or doubled without altering their shapes.

Doubling of numbers was standard technique in Egyptian arithmetic; so in measuring land areas the relations between the double-remen, the cubit, and the remen enabled areas (whether squares, rectangles, triangles, or other shapes) to be doubled and halved merely

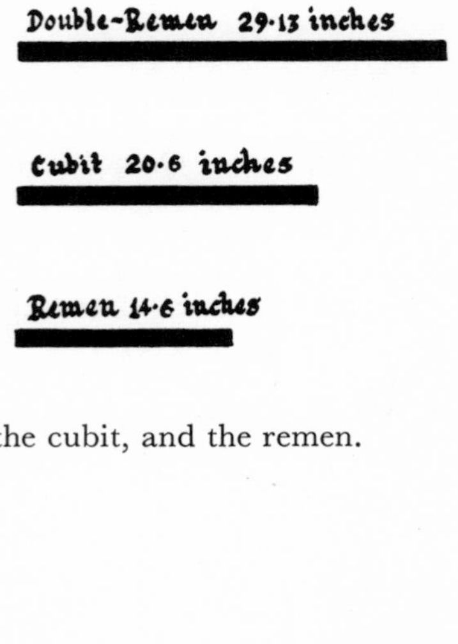

Double-Remen 29.13 inches

Cubit 20.6 inches

Remen 14.6 inches

FIGURE 20.1
The double-remen, the cubit, and the remen.

TABLE 20.1
Table of length and fractions of a cubit.

4 fingers = 1 palm				
7 palms = 1 cubit				
Fractions of a Cubit				
Palms	Cubits			
1	$\bar{7}$			
2	$\bar{4}$	$\overline{28}$ a		
3	$\bar{4}$	$\bar{7}$	$\overline{28}$	
4	$\bar{2}$	$\overline{14}$		
5	$\bar{2}$	$\bar{7}$	$\overline{14}$	
6	$\bar{2}$	$\bar{4}$	$\overline{14}$	$\overline{28}$

a From the RMP Recto, 2 ÷ 7.

by changing the units of measurement while preserving the proportions of the figures. Thus the relative lengths in Figure 20.1 are such
that a square on the side of the double-remen is double the area of a
square on the cubit, while a square on the side of the remen is half a
square on the cubit.

From a table such as Table 20.1 the scribe could have deduced from
(1 + 6) or (2 + 5) or (3 + 4) palms equaling one cubit, the equality
$\bar{2}$ $\bar{4}$ $\bar{7}$ $\overline{14}$ $\overline{28}$ = 1, which is what he gave in RMP 38. Further, since
$\bar{2}$ $\bar{4}$ $\bar{4}$ = 1, he had at once $\bar{7}$ $\overline{14}$ $\overline{28}$ = $\bar{4}$, an equality he frequently
used, as for example in RMP Problems 9, 10, 11, 12, 14, 24, and
elsewhere.

ARURA:
A unit of area not commonly met is the *arura*, equal to the area of a
square whose side was 100 royal cubits, and thus 10,000 square cubits.
The linear length of 100 royal cubits was called a *schoenia* in the time
of the Ptolemies, and this would be the length of the side of the square
of one arura. See hayt and khet. The measures quoted here are from
land measures in Greek papyri of the first century A.D.

KHET, SETAT:
The common unit for linear measures of land at the time of the RMP
was the *khet*, of 100 cubits. A *setat* was a square khet or 10,000 square
cubits. What was called a *cubit-strip* was a long rectangle of 100 cubits

by 1 cubit, or one khet by one cubit, 100 of which cubit-strips would make one setat.

HEKAT:

A hekat was a half-peck dry measure for barley, wheat, corn, and grain generally. It was thus $\frac{1}{8}$ bushel, or 4 quarts, or 8 pints dry measure. Chace gives it as 292.24 cubic inches,* while half a peck in British measure was 277.36 cubic inches, so that a hekat was slightly more than half a peck. Eisenlohr, Sethe, and Struve call it the *scheffel*, Gunn uses *gallon*, and Peet *bushel*. For stating the contents of larger grain vessels, the units might be double-hekats or more commonly quadruple-hekats, which would therefore be $\frac{1}{4}$ and $\frac{1}{2}$ of a modern bushel, respectively. For storage granaries an even larger unit was needed, and use was made then of "*100 quadruple hekat*" units. One cubic cubit contains 30 hekats of grain.

HINU:

The hinu was a smaller unit for grain, being one-tenth of a hekat.

KHAR:

This was two-thirds of a cubic cubit, or 20 hekats of grain. Or we can say 1 $\bar{2}$ khar make a cubic cubit.

RO:

The smallest named unit for grain was the ro, which was $\frac{1}{320}$ part of the hekat. It would be between a dessertspoon and a tablespoon full of grain. The only fractions of a hekat used were $\bar{2}, \bar{4}, \bar{8}, \overline{16}, \overline{32}, \overline{64}$, and these were written in a special way, quite unlike ordinary fractions. They were called *Horus-eye* fractions, and were used solely for grain (see Figure 20.2). Then $\overline{64}$ of a hekat contained 5 ro, and for any fraction of a hekat less than $\overline{64}$, ro and sometimes fractions of a ro had to be used. Horus was the son of Osiris, who was treacherously slain by his brother Seth. In revenge, Horus sought out his uncle and slew him, but in the fight lost an eye, the broken parts of which were later restored by the god Thoth. Isis was the mother of Horus, and the wife and sister of Osiris.

* A. B. Chace; L. Bull; H. P. Manning; and R. C. Archibald, *The Rhind Mathematical Papyrus*, Vol. 1, Mathematical Association of America, Oberlin, Ohio, 1927, p. 31.

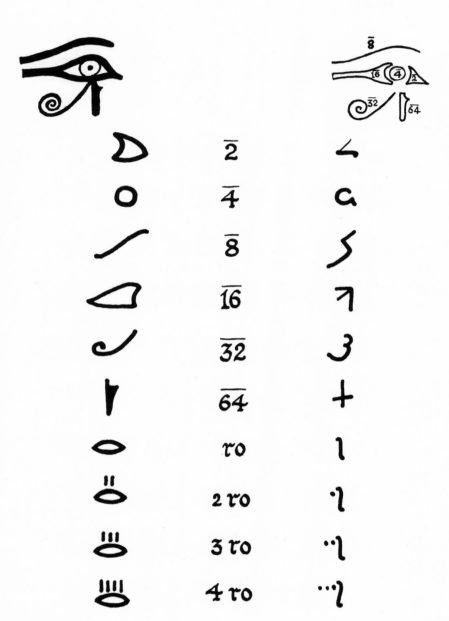

FIGURE 20.2
Horus-eye fractions.

DEBEN, SHATY:

In Problem 62 of the RMP, the value in *shaty* of one *deben* (weight) of gold is given as 12. Then one deben of silver is 6 shaty and one deben of lead is 3 shaty. According to Chace, the shaty was a seal (*sic*), the word representing a unit of value. It was not a coin. The deben was a weight of about 91 grams and consequently about 3.2 ounces avoirdupois. However Sloley* gives a deben as "the weight (1,470 grains) of the anklet of the same name, of which the tenth part was the *quedet*, the weight of the finger ring." Then at 7,000 grains per 1 lb. avoirdupois this would give a deben of 3.36 ounces avoirdupois.

SEKED:

The *seked* of a right pyramid is the inclination of any one of the four triangular faces to the horizontal plane of its base, and is measured as so many horizontal units per one vertical unit rise. It is thus a measure equivalent to our modern cotangent of the angle of slope. In general, the seked of a pyramid is a kind of fraction, given as so many palms horizontally for each cubit vertically, where 7 palms equal one cubit. The Egyptian word "seked" is thus related to our modern word "gradient."

PESU:

A pesu is a unit measuring the strength of beer, bread, or cakes, according to the amount of grain used. If one hekat of grain were used to make 10 loaves of bread, then their pesu was said to be 10; if one hekat made 15 loaves, then their pesu was 15. In the same way, if 1 hekat of grain was used to make 5 des-jugs of beer, then the beer was said to have a pesu of 5; if it made only 3 des-jugs, their pesu was 3. Thus the less the pesu, the stronger the beer (or bread), and the higher the pesu, the weaker the beer (or bread). The formula

$$\text{pesu} = \frac{\text{number of loaves or des-jugs}}{\text{number of hekats}}$$

expresses the relation. Authorities use the following terms to explain the meaning of pesu:

* *The Legacy of Egypt*, p. 176.

Chace   —Cooking ratio.
Vogel   —Backungzahl or Baking number.
Struve   —Backverhältnis or Baking proportion.
Van der Waerden—Baking Value.
Peet   —Cooking figure.

DES-JUG:
Struve says that one des-jug of beer is approximately half a liter, so
that it would therefore be about $7/8$ of a pint.

21 SQUARES AND SQUARE ROOTS

The squares of numbers, both integral and fractional, are quite often stated and calculated in the mathematical papyri, but square roots are far less common, and although stated, they are not calculated. There would have been no need for the Egyptians to devise a means of finding the square roots of perfect squares. These could have been read off from a table of the squares of integers. Such a table would have been very easy to construct, and indeed very probably was drawn up by the scribes. Similar tables involving the simpler fractions could equally well have been made by them, using ordinary Egyptian multiplication; and although no such tables have been preserved, if they were in fact made, they would have looked like Table 21.1. These can be read forwards and backwards equally well, and they would have been sufficient for all ordinary requirements. These tables could also be used to obtain good approximations to numbers not specifically listed. Let us suppose for instance that the square root of 40 is required. From the tables we read that the square root of 39 $\overline{16}$ is 6 $\overline{4}$, while the square root of 40 $\overline{9}$ is 6 $\overline{3}$. Then the square root of 40 lies somewhere between 6 $\overline{4}$ and 6 $\overline{3}$, and closer probably to 6 $\overline{3}$, so that this may be a sufficiently close approximation for the purpose at hand. But if it is not, we can proceed as follows with standard multiplication, but starting with the lesser value, 39 $\overline{16}$ being below 40, so that we can add smaller fractions to 6 $\overline{4}$, rather than subtract them from 6 $\overline{3}$. The value 6 $\overline{4}$ is less than the exact answer by 0.0746 to four decimal places in our notation. For the square of 6 $\overline{4}$ the working would be:

1		6	$\overline{4}$	
\2		12	$\overline{2}$	
\4		25		
$\overline{2}$		3	$\overline{8}$	
\$\overline{4}$		1	$\overline{2}$	$\overline{16}$
Totals 6	$\overline{4}$	39	$\overline{16}$.	

Now the table tells us that 6 $\overline{3}$ × 6 $\overline{3}$ is too great for the square root of 40, and therefore, since $\overline{4}$ $\overline{12}$ = $\overline{3}$ (by application of the G rule or

TABLE 21.1
Perfect squares as they might have been tabulated by the scribes.

No.	Square	No.	Square	No.	Square	No.	Square
1	1	$\bar{2}$	$\bar{4}$	$\bar{3}$	$\bar{9}$	$\bar{4}$	$\overline{16}$
2	4	1 $\bar{2}$	2 $\bar{4}$	1 $\bar{3}$	1 $\bar{3}$ $\bar{9}$	1 $\bar{4}$	1 $\bar{2}$ $\overline{16}$
3	9	2 $\bar{2}$	6 $\bar{4}$	2 $\bar{3}$	5 $\bar{3}$ $\bar{9}$	2 $\bar{4}$	5 $\overline{16}$
4	16	3 $\bar{2}$	12 $\bar{4}$	3 $\bar{3}$	11 $\bar{9}$	3 $\bar{4}$	10 $\bar{2}$ $\overline{16}$
5	25	4 $\bar{2}$	20 $\bar{4}$	4 $\bar{3}$	18 $\bar{3}$ $\bar{9}$	4 $\bar{4}$	18 $\overline{16}$
6	36	5 $\bar{2}$	30 $\bar{4}$	5 $\bar{3}$	28 $\bar{3}$ $\bar{9}$	5 $\bar{4}$	27 $\bar{2}$ $\overline{16}$
7	49	6 $\bar{2}$	42 $\bar{4}$	6 $\bar{3}$	40 $\bar{9}$	6 $\bar{4}$	39 $\overline{16}$
8	64	7 $\bar{2}$	56 $\bar{4}$	7 $\bar{3}$	53 $\bar{3}$ $\bar{9}$	7 $\bar{4}$	52 $\bar{2}$ $\overline{16}$
9	81	8 $\bar{2}$	72 $\bar{4}$	8 $\bar{3}$	69 $\bar{3}$ $\bar{9}$	8 $\bar{4}$	68 $\overline{16}$
10	100	9 $\bar{2}$	90 $\bar{4}$	9 $\bar{3}$	87 $\bar{9}$	9 $\bar{4}$	85 $\bar{2}$ $\overline{16}$

No.	Square	No.	Square	No.	Square	No.	Square
$\bar{2}$ $\bar{4}$	$\bar{2}$ $\overline{16}$	$\bar{3}$	$\bar{3}$ $\bar{9}$	$\bar{5}$	$\overline{25}$	$\bar{6}$	$\overline{36}$
1 $\bar{2}$ $\bar{4}$	3 $\overline{16}$	1 $\bar{3}$	2 $\bar{3}$ $\bar{9}$	1 $\bar{5}$	1 $\bar{3}$ $\overline{15}$ $\overline{25}$	1 $\bar{6}$	1 $\bar{3}$ $\overline{36}$
2 $\bar{2}$ $\bar{4}$	7 $\bar{2}$ $\overline{16}$	2 $\bar{3}$	7 $\bar{9}$	2 $\bar{5}$	4 $\bar{3}$ $\overline{10}$ $\overline{25}$ $\overline{30}$	2 $\bar{6}$	4 $\bar{3}$ $\overline{36}$
3 $\bar{2}$ $\bar{4}$	14 $\overline{16}$	3 $\bar{3}$	13 $\bar{3}$ $\bar{9}$	3 $\bar{5}$	10 $\bar{5}$ $\overline{25}$	3 $\bar{6}$	10 $\overline{36}$
4 $\bar{2}$ $\bar{4}$	22 $\bar{2}$ $\overline{16}$	4 $\bar{3}$	21 $\bar{3}$ $\bar{9}$	4 $\bar{5}$	17 $\bar{3}$ $\bar{5}$ $\overline{15}$ $\overline{25}$	4 $\bar{6}$	17 $\bar{3}$ $\overline{36}$
5 $\bar{2}$ $\bar{4}$	33 $\overline{16}$	5 $\bar{3}$	32 $\bar{9}$	5 $\bar{5}$	27 $\overline{25}$	5 $\bar{6}$	26 $\bar{3}$ $\overline{36}$
6 $\bar{2}$ $\bar{4}$	45 $\bar{2}$ $\overline{16}$	6 $\bar{3}$	44 $\bar{3}$ $\bar{9}$	6 $\bar{5}$	38 $\bar{3}$ $\overline{15}$ $\overline{25}$	6 $\bar{6}$	38 $\overline{36}$
7 $\bar{2}$ $\bar{4}$	60 $\overline{16}$	7 $\bar{3}$	58 $\bar{3}$ $\bar{9}$	7 $\bar{5}$	51 $\bar{3}$ $\overline{10}$ $\overline{30}$ $\overline{25}$	7 $\bar{6}$	51 $\bar{3}$ $\overline{36}$
8 $\bar{2}$ $\bar{4}$	76 $\bar{2}$ $\overline{16}$	8 $\bar{3}$	75 $\bar{9}$	8 $\bar{5}$	67 $\bar{5}$ $\overline{25}$	8 $\bar{6}$	66 $\bar{3}$ $\overline{36}$
9 $\bar{2}$ $\bar{4}$	95 $\overline{16}$	9 $\bar{3}$	93 $\bar{3}$ $\bar{9}$	9 $\bar{5}$	84 $\bar{3}$ $\bar{5}$ $\overline{15}$ $\overline{25}$	9 $\bar{6}$	84 $\overline{36}$

No.	Square	No.	Square	No.	Square
$\bar{7}$	$\overline{49}$	$\bar{8}$	$\overline{64}$	$\bar{9}$	$\overline{81}$
1 $\bar{7}$	1 $\bar{4}$ $\overline{28}$ $\overline{49}$	1 $\bar{8}$	1 $\bar{4}$ $\overline{64}$	1 $\bar{9}$	1 $\bar{6}$ $\overline{18}$ $\overline{81}$
2 $\bar{7}$	4 $\bar{2}$ $\overline{14}$ $\overline{49}$	2 $\bar{8}$	4 $\bar{2}$ $\overline{64}$	2 $\bar{9}$	4 $\bar{3}$ $\bar{9}$ $\overline{81}$
3 $\bar{7}$	9 $\bar{2}$ $\bar{4}$ $\overline{14}$ $\overline{28}$ $\overline{49}$	3 $\bar{8}$	9 $\bar{2}$ $\bar{4}$ $\overline{64}$	3 $\bar{9}$	9 $\bar{3}$ $\overline{81}$
4 $\bar{7}$	17 $\bar{7}$ $\overline{49}$	4 $\bar{8}$	17 $\overline{64}$	4 $\bar{9}$	16 $\bar{3}$ $\bar{6}$ $\overline{18}$ $\overline{81}$
5 $\bar{7}$	26 $\bar{4}$ $\bar{7}$ $\overline{28}$ $\overline{49}$	5 $\bar{8}$	26 $\bar{4}$ $\overline{64}$	5 $\bar{9}$	26 $\bar{9}$ $\overline{81}$
6 $\bar{7}$	37 $\bar{2}$ $\bar{7}$ $\overline{14}$ $\overline{49}$	6 $\bar{8}$	37 $\bar{2}$ $\overline{64}$	6 $\bar{9}$	37 $\bar{3}$ $\overline{81}$
7 $\bar{7}$	51 $\overline{49}$	7 $\bar{8}$	50 $\bar{2}$ $\bar{4}$ $\overline{64}$	7 $\bar{9}$	50 $\bar{2}$ $\overline{18}$ $\overline{81}$
8 $\bar{7}$	66 $\bar{4}$ $\overline{28}$ $\overline{49}$	8 $\bar{8}$	66 $\overline{64}$	8 $\bar{9}$	65 $\bar{3}$ $\bar{9}$ $\overline{81}$
9 $\bar{7}$	83 $\bar{2}$ $\overline{14}$ $\overline{49}$	9 $\bar{8}$	83 $\bar{4}$ $\overline{64}$	9 $\bar{9}$	83 $\overline{81}$

gen. $(1, 3)$), the addition of $\overline{12}$ to $6\ \overline{4}$ giving $6\ \overline{4}\ \overline{12}$ as a closer approximation would be too great. Then the choice is to be made from the smaller unit fractions, $\overline{13}, \overline{14}, \overline{15}, \overline{16}, \overline{17}, \ldots$. Here a certain judgment is required, in which past experience would help. It is also of some practical importance to keep the multiplications as simple as possible. Suppose we choose $\overline{16}$. The multiplication is:

1			6	$\overline{4}$	$\overline{16}$		
\2			12	$\overline{2}$	$\overline{8}$		
\4			25	$\overline{4}$			
$\overline{2}$			3	$\overline{8}$	$\overline{32}$		
\$\overline{4}$			1	$\overline{2}$	$\overline{16}$	$\overline{64}$	
$\overline{8}$				$\overline{2}$	$\overline{4}$	$\overline{32}$	$\overline{128}$
\$\overline{16}$				$\overline{4}$	$\overline{8}$	$\overline{64}$	$\overline{256}$

Totals $6\ \overline{4}\ \overline{16}$ $39\ \overline{4}\ \overline{4}\ \overline{8}\ \overline{8}\ \overline{16}\ \overline{64}\ \overline{64}\ \overline{256}$

                                         $39\ \overline{2}\ \overline{4}\ \overline{16}\ \overline{32}\ \overline{256}$.

Then $6\ \overline{4}\ \overline{16}$ is a closer approximation and may have been sufficient for the scribe's purposes, although it is still too small by 0.0121.

In KP LV, 4, lines 39, 40, Griffith translates, "*make thou a corner (square root of 16) as 4*," where the pertinent hieroglyphs are: ⟨hieroglyphs⟩ A corner thou make.*

In the Berlin Papyrus 6619, Schack-Schackenburg translates as, "Nimm die Quadratwurzel daraus ($1\ \overline{2}\ \overline{16}$), das giebt $1\ \overline{4}$" (Take the square root of $1\ \overline{2}\ \overline{16}$, it is $1\ \overline{4}$), the line he displays as:†

⟨hieroglyphs⟩

In the Moscow Papyrus, Problem 6, Struve's translation reads, "Berechne du seinen Winkel (Quadratwurzel 16). Es entsteht 4." (Calculate thou its angle (square root 16). Result 4.)

Peet notes in his *Mathematics in Ancient Egypt* that in the Berlin Papyrus 6619 the square root of $6\ \overline{4}$ is also correctly given by the scribe. See the table of squares (Table 21.1) and Chapter 14.

* Read the hieroglyphs from right to left.
† Read left to right.

R. C. Archibald* notes that in a Greek papyrus dating from the second century (P11529, Berlin Museum), in the first of five problems dealing with areas of fields in arurae, the square root of 164 is found in the form 12 $\overline{3}$ $\overline{15}$ $\overline{26}$ $\overline{32}$, which is a good approximation. If $\overline{26}$ were an error for $\overline{24}$ the approximation would indeed be remarkable.

* *The Rhind Mathematical Papyrus*, Vol. 1, p. 176.

22 THE REISNER PAPYRI: THE SUPERFICIAL CUBIT AND SCALES OF NOTATION

The Reisner Papyri were found by Dr. George Reisner in 1904 at Nagᶜed Deir in Upper Egypt, during excavations being carried out for Harvard University and the Boston Museum of Fine Arts. The four badly worm-eaten rolls were found lying on a wooden coffin. Dr. H. Ibscher, the director of the papyrus collection of the Akademie der Wissenschaften (Berlin), unrolled and restored RP 1, but the remaining rolls remained in Berlin until the end of World War II. These papyri are for the most part the official registers of dockyard workshops, Nagᶜed Deir being a necropolis for the town of This. Originally the whole papyrus was about $11\frac{1}{2}$ feet long and about a foot high. The editor and principal translator,* Dr. W. K. Simpson, concludes that the papyri date from the reign of Sesostris I of the Twelfth Dynasty, from which I conclude their approximate date to be 1880 B.C. In the first two volumes on the Reisner Papyri,† Simpson makes 17 divisions as shown in Table 22.1.

We are concerned only with the sections marked G, H, and I. Section G is an analysis of the volume calculations of the rectangular excavations planned for the temple to be built and the determination of the number of enlistees, or workmen, required to make them. Division by 10 suggests that 10 cubic cubits is the unit of volume expected of each enlistee per day. Section H is an analysis of the calculations for the volumes of blocks of stone from the storehouse; while Section I, which is the "least clear of the three,"‡ probably deals with the volumes of rectangular walls or floors, as I 7 suggests, length 52 cubits, width 3 cubits, and thickness (or depth) $\overline{4}$ cubit, perhaps a wall or pathway. No workings for the clerk's calculations are shown, but

---

* Earlier Dr. Alan H. Gardiner and Paul C. Smither (d. 1943 at age 29) made some translations.
† Papyrus Reisner I (1963), II (1965), III (1969), W. K. Simpson, Boston Museum of Fine Arts.
‡ Simpson.

TABLE 22.1
W. K. Simpson's seventeen divisions of the Reisner Papyri.

Sections A, B.	Rosters of enlistees.
Section C.	List of foremen and laborers.
Section D.	List of enlistees going upstream.
Section E.	Names of foremen and their crews.
Section F.	Lists of foremen, clerks, and crews.
Section G.	Calculations of temple excavations and enlistees needed.
Section H.	Calculations of volumes of stone blocks from the storehouse.
Section I.	Calculations, floor plans, walls, trenches, corridors.
Section J.	Totals of enlistees from the above sections.
Section K.	Divisions by 10 for enlistees from the above sections.
Section L.	Account of cargo, hides, cattle, fish, oil, pigeons, fowls.
Section M.	Various totals.
Section N.	List of officials, mostly women.
Section O.	Like a balance sheet, 6 long columns of numbers.
Section P.	A second roster of names, men.
Section Q.	Contains many large numbers, such as 40,566, 39,548.

certainly they must have been done on some sort of papyritic writing pad, from which the answers were transferred to the RP lists. What would we not give to see *that* piece of papyrus! I would hazard the guess that as each block of stone was brought to the site, an official measurer would call out the dimensions for the clerk to write down. While waiting for the next to arrive, he would do the multiplication, which would be entered on the appropriate line. Multiplications and divisions would also have been done for the heading "units," where there was more than one unit of the same size, and the heading "enlistees," for the results of the divisions by 10.

The measuring reed used, whatever its overall length, must have been marked in cubits, palms, and fingers, as well as in halves, thirds, and quarters of cubits. It was probably 6 cubits long and may have looked like the one shown in Figure 22.1.

Table 22.4 is one that could have been constructed by the scribe in order to ease the work of dividing by 10. This table may be constructed by ordinary division, or from the table preceding Problems 1–6 of the RMP. The entries of Table 22.4 should be compared with the scribal work collected in Tables 22.2 (correct calculations) and 22.3 (scribal

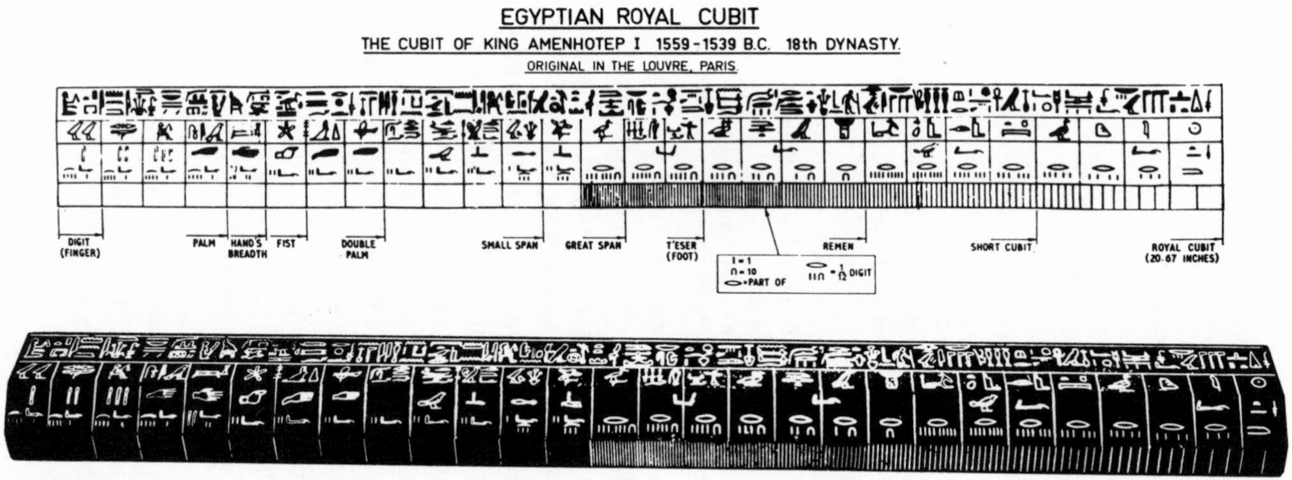

FIGURE 22.1

A possible division of an Egyptian measuring-rod. Courtesy M. J. Puttock B.Sc., National Standards Laboratory, C.S.I.R.O., Sydney, Australia.

errors and approximations). Thus, in line G10 (Table 22.3B) and line I7 (Table 22.2), the clerk has written the approximation $39 \div 10 = 4$ in his calculations for the Eastern Chapel. But he need not have made this approximation, for a glance at Table 22.4 gives immediately

$$39 \div 10 = (30 + 9) \div 10$$
$$= 3\ \bar{2}\ \bar{3}\ \bar{15}.$$

TABLE 22.2
Calculations by the scribe of the Reisner Papyri which are quite correct.

Line	l.	b.	d.	Units	Volume	Enlistees	Detail
G 5	3	2	2	1	12	1 $\bar{5}$	
G 6, H 32	8	5	$\bar{4}$	1	10	1	Eastern Chapel.
G 14	8	3	$\bar{3}$	1	8	$\bar{2}\ \bar{4}\ \bar{20}$	
G 15	6	4	2	1	48	4 $\bar{2}\ \bar{4}\ \bar{20}$	
G 16	4	2	2	1	16	1 $\bar{2}\ \bar{10}$	
G 17, H 33	4	4	2	2	64	6 $\bar{4}\ \bar{10}\ \bar{20}$	Footings.
G 18, H 34	3	3	2	2	36	3 $\bar{2}\ \bar{10}$	Footings.
H 31	15	5	$\bar{4}$	1	18 $\bar{2}\ \bar{4}$		Great Chamber.
H 7	3 $\bar{4}$	1 $\bar{2}$	$\bar{2}$	2	4 $\bar{2}\ \bar{4}\ \bar{8}$		
H 8	2 $\bar{2}$	1 $\bar{2}$	$\bar{2}$	1	1 $\bar{2}\ \bar{4}\ \bar{8}$		
H 9	4 $\bar{2}$	1 $\bar{2}$	1	2	13 $\bar{2}$		
H 11	3c 1p	1	1	1	3c 1p		
H 17	4	1c 3p	1	4	22c 6p		
H 25	1 $\bar{2}$	$\bar{2}\ \bar{4}$	1	2	2 $\bar{4}$		
H 26	2 $\bar{2}$	$\bar{2}\ \bar{4}$	1	2	3 $\bar{2}\ \bar{4}$		
H 27	3 $\bar{2}$	1 $\bar{2}$	1 $\bar{2}$	2	¡5 $\bar{2}\ \bar{4}$		
H 30	12	5	$\bar{4}$	1	15		Great Chamber.
I 2	12	5	$\bar{2}$	1	30		Great Chamber.
I 3	15	5	$\bar{2}$	1	37 $\bar{2}$		August Chamber.
I 4	8	5	$\bar{2}$	1	20		Eastern Chapel.
I 5	18	11	$\bar{\bar{3}}$	1	132		
I 6	32	4	$\bar{4}$	1	32		Western.
I 7, G 10	52	3	$\bar{4}$	1	39		Eastern.
I 8	24	5p	$\bar{2}$	1	8c 4p		
I 9	26	2	5p	1	111c 3p		Carrying srft.
I 10	20	5	5p	1	71c 3p		Carrying srft.
I 12	27	7	2	1	378		Loosening brick clay.
I 13	8	7	2	1	112		Water from a field.
I 14	1 $\bar{2}$	1 $\bar{2}$	2	2	9		For tower.
I 15	2 $\bar{2}$	1 $\bar{2}$	1 $\bar{2}$	2	11 $\bar{4}$		For tower.
I 20	8	6	1	1	48		

TABLE 22.3

(A, B, C, D, E) Calculations which appear to be errors, major and minor, with restorations.

A. Calculations by the Clerk Correct Except for Simple or Obvious Scribal Errors.

Line	l.	b.	d.	Units	Volume	Enlistees	Detail
G 8	35	11	$\bar{2}$	1	192 $\bar{2}$	19 $\bar{2}$ for 19 $\bar{4}$	
H 35	8	5 for 9	$\bar{4}$	1	18		
I 16	3 $\bar{2}$	2 $\bar{2}$	1 $\bar{2}$	2	25 $\bar{4}$ for 26 $\bar{4}$		For the tower.
I 17	4	2 $\bar{2}$	1 $\bar{2}$	2	36 for 30		For the tower.
I 18	10	5 $\bar{2}$	$\bar{4}$ for 1	1	55		Brick clay.

B. Errors, Misreadings, or Approximations.

Line	l.	b.	d.	Units	Volume	Enlistees
G 7	3	2 $\bar{2}$	4	1	2 $\bar{2}$ for 1 $\bar{2}$ $\bar{4}$ $\bar{8}$	5 $\overline{20}$ correct for 2 $\bar{2}$.
G 9	13	11	1 $\bar{2}$	1	214 $\bar{2}$	21 $\bar{2}$ for 21 $\bar{4}$ 5.
G 10	52	3	$\bar{4}$	1	39	4 for 3($\bar{2}$ $\bar{3}$ $\overline{15}$).
G 11	32	4	$\bar{2}$	1	85 for 64	8 $\bar{2}$ correct for 85.
G 12	3 $\bar{2}$	2	$\bar{3}$	1	4 $\bar{3}$	$\bar{2}$ for ($\bar{3}$ $\overline{10}$ $\overline{30}$).
G 13	10 $\bar{2}$	8 $\bar{2}$	$\bar{3}$	1	27 for 29 $\bar{2}$ $\bar{4}$	2 $\bar{2}$ 5 correct for 27.
H 20	3c 3p	1c 3p	1	1	5c	5c for 4c 6p ($\bar{4}$ $\overline{28}$)p.

C. Minor Errors.

Line	l.	b.	d.	Units	Volume
H 10	4c 1p	1 $\bar{2}$	$\bar{2}$	1	(3c 2f) for (3c 3f).
H 13	2c 3p	2c 3p	$\bar{3}$	1	(3c 6p 1 $\bar{3}$f) for (3c 6p 2f)($\overline{14}$ $\overline{42}$f).
H 14	2c 2f	1 $\bar{2}$	1c 1p 1f	1	(3c 4p 1 $\bar{2}$f) for (3c 4p 2f)($\bar{2}$ $\overline{28}$f).
H 15	1c 5p	1 $\bar{2}$	5p	2	(3c 5p) for (3c 4p 2f)($\bar{2}$ $\bar{4}$ $\overline{14}$ $\overline{28}$f).
H 18	3c 2p	1c 2p	6p	1	(3c 3p 2 $\bar{3}$f) for (3c 4p 1 $\bar{3}$f)($\overline{15}$f).
H 22	1 $\bar{\bar{3}}$	1c 3p	1	1	(2c 2p 2f) for (2c 2p 2 $\bar{3}$f).
H 24	3c 5p	1c 2p	6p	1	(4c 2f) for (4c 2 $\bar{3}$f)(5 $\overline{15}$) approx.
H 28	4c 4p	1c 5p	1c 2p	2	(20c 1p 1 $\bar{2}$f) for (20c 1p $\bar{4}$f) approx.

D. Major Errors.

Line	l.	b.	d.	Units	Volume
H 16	2c 3p	1c 4p	5p 2f	1	(2c 5p 2 $\bar{2}$f) for (2c 6p 3 $\bar{2}$f) approx.
H 19	3c 5p 2f	1c 3p	1	1	(4c 2p 3f) for (5c 2p 3 $\bar{4}$f)(7 $\overline{28}$).
H 21	1c 5p	1c 3p	1	1	(2c 4p 1...f) for (2c 3p $\bar{2}$ $\overline{14}$f).
H 23	4	1c 6p	6p	1	(4c 1f) for (6c 2p 2 $\bar{4}$f)28.

E. Possible Restorations.

Line l.	b.	d.	Units	Volume	Restoration
H 2 2c 5p	6p	[ ]	2	4c 1 $\bar{3}$f	d = 6p 1f (probably too great).
H 3 2 $\bar{4}$	6p	[ ]	1	1c 2p 2 . . . f	d = 5p (gives v = 1c 2p 2 $\bar{2}$ $\overline{14}$f).
H 4 2 $\bar{4}$	6p	[ ]	[ ]	1 . . . 2p 2 . . . f	d = 5p (units = 1, as above).
H 5 2 $\bar{4}$	6p	[ ]	1	1c 3p 1 $\bar{2}$f	d = 5p 2f (probably too great)
H 6 4 $\bar{4}$	[ ]	[ ]	2	6 $\bar{2}$ $\bar{4}$	b = 6p

TABLE 22.4
Table for dividing by 10 for enlistees.

The Number	Divided by 10	Alternatively
$\bar{\bar{3}}$	$\overline{15}$	
$\bar{2}$	$\overline{20}$	
$\bar{3}$	$\overline{30}$	
$\bar{4}$	$\overline{40}$	
1	$\overline{10}$	
2	$\bar{5}$	
3	$\bar{5}$ $\overline{10}$	
4	$\bar{3}$ $\overline{15}$	($\bar{4}$ $\overline{10}$ $\overline{20}$)
5	$\bar{2}$	
6	$\bar{2}$ $\overline{10}$	($\bar{3}$ $\bar{6}$ $\overline{10}$)
7	$\bar{2}$ $\bar{5}$	($\bar{\bar{3}}$ $\overline{30}$)
8	$\bar{2}$ $\bar{5}$ $\overline{10}$	($\bar{2}$ $\bar{4}$ $\overline{20}$)
9	$\bar{2}$ $\bar{3}$ $\overline{15}$	($\bar{\bar{3}}$ $\bar{6}$ $\overline{15}$)
10	1	
20	2	
30	3	

My interpretation of Simpson's translation of Reisner Papyri I and II leads me to conclude that the chief overseer of the dockyard appointed a skilled scribe to instruct the tally clerk in multiplication and division for the records of the workshop. The scribe's teaching for integers and the simple fractions of a cubit like $\bar{\bar{3}}$, $\bar{2}$, $\bar{3}$, $\bar{4}$ was well done, because of the 65 entries made by the clerk, there were 11 that contained only integral values of cubits, and for these the clerk's arithmetic is 100 percent accurate. As an example, we take line G 18, footings (Table 22.2):

length	breadth	depth	units	volume	enlistees
3 c	3 c	2 c	2	36 c	3 $\bar{2}$ $\overline{10}$

The calculations not shown by the clerk must have been,

\1 c	3 c
\2 c	6 cc
Totals 3 c	9 cc
1 c	9 cc
Totals \2 c	18 ccc
1	18 ccc
Totals \2	36 ccc volume.

36 ccc ÷ 10: \ 1	10
\ 2	20
\ $\bar{2}$	5
\$\overline{10}$	1
Totals 3 $\bar{2}$ $\overline{10}$	36 enlistees. (See Table 22.4)

Of course, some of the simpler steps may have been done mentally, and almost certainly, Egyptian equivalents of our abbreviations c, cc, and ccc were not included in the calculations, although they must have been mentally noted.

In those lines where *fractions* of cubits were included, only 3 errors occurred, so that in the 31 of these calculations, the clerk was 90 percent accurate. Here is an example from H 27 (Table 22.2):

length	breadth	depth	units	volume
3 $\bar{2}$ c	1 $\bar{2}$ c	1 $\bar{2}$ c	2	15 $\bar{2}$ $\bar{4}$

The calculations not shown by the clerk may have been,

\1 c	3 $\bar{2}$ c
\$\bar{2}$ c	1 $\bar{2}$ $\bar{4}$ cc
Totals 1 $\bar{2}$ c	5 $\bar{4}$ cc.
\1 c	5 $\bar{4}$ ccc
\$\bar{2}$ c	2 $\bar{2}$ $\bar{8}$ ccc
Totals 1 $\bar{2}$ c	7 $\bar{2}$ $\bar{4}$ $\bar{8}$ ccc.
1	7 $\bar{2}$ $\bar{4}$ $\bar{8}$ ccc
Totals \2	15 $\bar{2}$ $\bar{4}$ ccc volume.

TABLE 22.5
Table of fingers, palms, and cubits.

4 fingers =	1 palm				
7 palms =	1 cubit				
1p =	$\overline{7}$				c
2p =	$\overline{4}$	$\overline{28}$			c
3p =	$\overline{4}$	$\overline{7}$	$\overline{28}$		c
4p =	$\overline{2}$	$\overline{14}$			c
5p =	$\overline{2}$	$\overline{7}$	$\overline{14}$		c
6p =	$\overline{2}$	$\overline{4}$	$\overline{14}$	$\overline{28}$	c

There are 11 entries that include measures in cubits and palms but *no fractions*. Now here the clerk began to find a little difficulty, because the palms must be expressed as fractions of a cubit, and so a table of fractions of a cubit would need to be prepared and handy for reference. The table would have been something like Table 22.5. See also Table 20.1.

Then let us look at line H 24, Table 22.3C, which falls in this category:

length	breadth	depth	units	volume
3 c 5 p	1 c 2 p	6 p	1	4 c 2 f

The first multiplication here is $(3\ \overline{2}\ \overline{7}\ \overline{14}) \times (1\ \overline{4}\ \overline{28})$, and we set it down as it must have been done by the clerk.

\ 1		3	$\overline{2}$	$\overline{7}$	$\overline{14}$		
$\overline{2}$		1	$\overline{2}$	$\overline{4}$	$\overline{14}$	$\overline{28}$	
\ $\overline{4}$			$\overline{2}$	$\overline{4}$	$\overline{8}$	$\overline{28}$	$\overline{56}$
\ $\overline{28}$			$\overline{14}$	$\overline{28}$	$\overline{56}$	$\overline{196}$	$\overline{392}$
Totals 1 $\overline{4}$ $\overline{28}$ c		4	$\overline{2}$	$\overline{7}$	$\overline{8}$	$\overline{196}$	$\overline{392}$ cc.

Now we have not shown how the clerk added the thirteen fractions $(\overline{2}\ \overline{7}\ \overline{14})\ (\overline{2}\ \overline{4}\ \overline{8}\ \overline{28}\ \overline{56})\ (\overline{14}\ \overline{28}\ \overline{56}\ \overline{196}\ \overline{392})$, and there remains further the multiplication by $(\overline{2}\ \overline{4}\ \overline{14}\ \overline{28})$ for the depth 6 palms, which will produce more than 20 fractions to be added! Clearly, a shorter method had to be found by the scribe and then explained to the clerk, a good calculator but not necessarily a mathematician. Perhaps the scribe might have thought to show how each dimension could be expressed in palms, and thus the multiplication for line H 24 (Table 22.3C), $26 \times 9 \times 6$, would give the volume in cubic palms, but the product would need to be divided by 343 in order to obtain cubic cubits. Thus,

\1		26
2		52
4		104
\8		208
Totals	9	234.
1		234
\2		468
\4		936
Totals	6	1404 cubic palms.

Divide by 343:

1	343				
2	686				
\ 4	1372				
$\bar{2}$	171	$\bar{2}$			
$\bar{4}$	85	$\bar{2}$	$\bar{4}$		
$\bar{8}$	42	$\bar{2}$	$\bar{4}$	$\bar{8}$	
\$\overline{16}$	21	$\bar{4}$	$\bar{8}$	$\overline{16}$	
\$\overline{32}$	10	$\bar{2}$	$\bar{8}$	$\overline{16}$	$\overline{32}$

Totals 4 $\overline{16}$ $\overline{32}$ 1404 $\bar{8}$ $\overline{32}$.   ($\bar{8}$ $\overline{32}$ in excess)

This is inaccurate, and furthermore, the scribe still had to find something simpler for the clerk. What the scribe found I think we can locate in line H 11 (Table 22.2):

length	breadth	depth	units	volume
3 c 1 p	1	1	1	3 c 1 p

The volume, here found mentally, clearly means 3 cubic cubits and one seventh of a cubic cubit, although it is written as 3 cubits 1 palm. One seventh of a cubic cubit would be a flat rectangular prism, 1 cubit by 1 cubit by 1 palm, and from the analogy of the modern "superficial foot" in measuring timber, I will call this the Egyptian "superficial cubit." This I think is quite justified, since in every one of the RP calculations, the volume is stated simply in cubits, palms, and fingers, without any suggestion of square cubits or palms, nor of cubic cubits or palms. And this simplified method of stating the volume or cubic contents allows of a much easier way of doing the required multiplications, which I am sure the scribe invented and explained to the clerk, although I have no direct evidence to prove it, working only with what is to be deduced from the multiplications before us in the Reisner Papyri.

We revert then to H 24 (Table 22.3C), where the ordinary standard Egyptian techniques left us in a maze of complicated multiplications of almost unmanageable fractions. This is what I think the scribe devised for this kind of multiplication. I repeat line H 24 here for convenience:

length	breadth	depth	units	volume
3 c 5 p	1 c 2 p	6 p	1	4 c 2 f.

Then,

	c	p		c	p	
				3	5	
\1				3	5	
\		2		6	10	
Totals	1	2		4	5	3.
				4	5	3
\		6		24	30	18
Totals		6		4 0	4	4.

The answer is 4 c 0 p and (4 4); amounts which the clerk wrote (in error I think) as 2 f, though it is a rough approximation. What the scribe has invented is what we call in modern textbooks *scales of notation*. In this case the additions were made in the *scale of 7*, simply because there are 7 palms in a cubit. Thus in the first addition, since $10 = 7 + 3$, put down 3 and carry 1; then $1 + 6 + 5 = 12$, which is 7 + 5, so put down 5 and carry 1; then 1 + 3 = 4. In the second addition, 18 = 2 × 7 + 4, put down 4 and carry 2; 2 + 30 = 32 = 4 × 7 + 4, put down 4 and carry 4; 4 + 24 = 28 = 4 × 7 + 0, put down 0 and carry 4. Of course zero, which had not yet been invented, was not written down by the scribe or clerk; in the papyri, a blank space indicates zero.

This technique works splendidly with every calculation of the RP, but it was new to the clerk, who got into many difficulties, particularly with H 14 (Table 22.3C), H 16, and H 19 (Table 22.3D), where the measurer was so meticulous as to include fingers in his linear measurements. But we should not be too critical of him, for he was seldom very far from the correct answer. It will help us to understand the scribe's thought processes if the multiplication is rewritten to include the units c and p for cubits and palms, so that cc and pp mean square cubits and square palms, and ccc and ppp mean cubic cubits and cubic palms. On this basis, cp means an area one cubit by one palm, and ccp means one square cubit by one palm, and thus a superficial cubit. Again, we have 7 cp = 1 cc, 7 pp = 1 cp, 7 ppp = 1 cpp, and so on, because we are working in the scale of 7. Then here is the working of H 24, repeated with these refinements.

		3 c	5 p		
\1 c		3 cc	5 cp		
\	2 p		6 cp	10 pp	
Totals		4 cc	5 cp	3 pp	
\	6 p		24 ccp	30 cpp	18 ppp
Totals		4 ccc	0 ccp	4 cpp	4 ppp.

Thus the volume is 4 cubic cubits, 0 superficial cubits, plus the small volumes represented by 4 cpp and 4 ppp, which the clerk recorded as 2 f, presumably as an approximation. If, by analogy with ccp for a superficial cubit (written simply as p by the clerk), ccf was written by him simply as f, then we should expect for (4 cpp + 4 ppp), 2 ccf + ($\overline{2}$ $\overline{14}$ $\overline{28}$ $\overline{196}$) ccf,* which latter he may have regarded as negligible,

* 4 cpp = ($\overline{2}$ $\overline{14}$) ccp = (2 $\overline{4}$ $\overline{28}$) ccf. And 4 ppp = ($\overline{2}$ $\overline{14}$) cpp = ($\overline{14}$ $\overline{98}$) ccp = ($\overline{4}$ $\overline{28}$ $\overline{28}$ $\overline{196}$) ccf = ($\overline{4}$ $\overline{14}$ $\overline{196}$) ccf. See table 22.5.

and thus wrote 2 f as an abbreviation meaning 2 ccf. The foregoing calculation and all the other calculations provide strong evidence that this technique, with perhaps some hieratic signs to identify ccc, ccp, cpf, ppf, etc., was indeed used by the clerk on his memo pad, and all the answers confirm the conclusion that, in stating the volume or cubical contents, c meant cubic cubits, and p meant a superficial cubit, that is, an area of one square cubit by one palm in thickness. I am equally well convinced that f or fb for fingerbreadths in the volume column meant a quarter of a superficial cubit, or an area of one square cubit by one fingerbreadth, since a fingerbreadth is a quarter of a palm. The evidence available is not so strong for this latter, because none of the lines involving fingerbreadths is entirely free of an error of some kind. Indeed where fractions, cubits, and palms are involved, the clerk is in error (even though very slightly) in 44 percent of the cases; but where fractions, cubits, palms, *and* fingerbreadths are concerned, he is wrong in *every* case. But despite this, what we may call the circumstantial evidence clearly points to a cubical content measure of one-quarter of a superficial cubit, which is itself one-seventh of a cubic cubit, and the scribal designation of this unit is f. Very little thought is necessary to see that we cannot call it a *superficial palm*, meaning one square palm by a fingerbreadth, for this volume would be one forty-ninth of the volume the clerk indicates by f. We will just have to remember that f means a quarter of a superficial cubit, as the clerk must have had to do. Thus,

$$1 \text{ cubic cubit } = 7 \text{ superficial cubits,}$$
$$\text{or } 1 \text{ c } = 7 \text{ p.}$$
$$1 \text{ superficial cubit } = 4 \text{ quarters of superficial cubits,}$$
$$\text{or } 1 \text{ p } = 4 \text{ f.}$$

The analogy with the more familiar Egyptian table of length must have been a very important factor in devising and adopting this method of calculating volumes, at least in the dockyard workshops of Nag^{c}ed Deir.

I conclude this chapter by attempting to show how the clerk calculated H 14 of Table 22.3C, in which the measurer included cubits, palms, fingerbreadths, *and* fractions; the lot! We note also that when adding, besides working in the scale of 7 for the conversion of p to c, we must work in the scale of 4 for the conversion of f to p. It could well be that it was in these conversions that the clerk became confused. All that is shown for H 14 is:

length	breadth	depth	units	volume
2 c 2 f	1 $\bar{2}$	1 c 1 p 1 f	1	3 c 4 p 1 $\bar{2}$ f

The working would have been:

	c	p	f		c	p	f		
					2		2		
Totals ＼1 $\bar{2}$					3 cc		3 cf		
					3 cc		3 cf		
	＼1 c				3 ccc		3 ccf		
	＼	1 p				3 ccp		3 cpf	
	＼		1 f			3 ccf			3 cff
Totals	1 c	1 p	1 f		3 ccc	4 ccp	2 ccf	3 cpf	3 cff.

The volume is therefore (3 c 4 p 2 f) plus some smaller fractions represented by (3 cpf 3 cff). The clerk had somehow made an error, where he had 1 $\bar{2}$ f for the finger column, but we can perhaps find out where the $\bar{2}$ came from. Thus,

$$
\begin{aligned}
3 \text{ cpf } 3 \text{ cff} &= 3 \text{ cpf } (\bar{2} \quad \bar{4}) \text{ cpf} & (3 \div 4) \\
&= (\bar{4} \quad \bar{7} \quad \overline{28}) \text{ ccf } (\overline{14} \quad \overline{28}) \text{ ccf} & (3 \div 7, \text{ tables}) \\
&= (\bar{4} \quad \bar{7} \quad \overline{14} \quad \overline{28} \quad \overline{28}) \text{ ccf} & (\text{rearranging}) \\
&= (\bar{4} \quad \bar{7} \quad \overline{14} \quad \overline{14}) \text{ ccf} & (\text{gen. } (1, 1)) \\
&= (\bar{4} \quad \bar{7} \quad \bar{7}) \text{ ccf} & (\text{gen. } (1, 1)) \\
&= (\bar{4} \quad \bar{4} \quad \overline{28}) \text{ ccf} & (\text{Recto, } 2 \div 7) \\
&= \bar{2} \quad \overline{28} \text{ ccf.} & (\text{gen. } (1, 1))
\end{aligned}
$$

The correct answer is therefore 3 c 4 p 2 $\bar{2}$ $\overline{28}$ f, for which the clerk had 3 c 4 p 1 $\bar{2}$ f.

If making such calculations as these was standard procedure in the building of a temple, and if the distribution of salary or rations to the overseers and workers was made in a manner similar to that shown in

Chapter 11 for the temple of Illahun, then what must have been the colossal mass of arithmetical calculation demanded by the building of the Great Pyramid or any of the other immense structures and monuments of ancient Egypt?

APPENDIX 1
THE NATURE OF PROOF

Those historians who have adversely criticized the mathematics of the ancient Egyptians confine themselves generally to the lack of formal proof in the Egyptian methods and to the apparent absence of what they call "the scientific attitude of mind" in the Egyptians' treatment of mathematical problems. Thus these commentators have written,

The fact that the Egyptians had evolved no better means of stating a formula than that of giving three or four examples of its use is hardly a tribute to the scientific nature of their mathematics.

The table [of the RMP Recto] is in itself a monument to the lack of the scientific attitude of mind in the Egyptians.

That they did not reach the conception of scientific mathematics and its dependence on cogent *a priori* demonstration is merely another instance of the vast debt which the world owes to the Greeks.   PEET*

There are no theorems (in the RMP) properly so called; everything is stated in the form of problems, not in general terms, but in distinct numbers.   JOURDAIN†

All available texts point to an Egyptian mathematics of rather primitive standards.   STRUIK‡

Egyptian geometry is not a science in the Greek sense of the word, but merely *applied arithmetic*.

The Greeks may also have taken from the Egyptians the rules for the determination of areas and volumes. But for the Greeks, such rules did not constitute mathematics; they merely led them to ask; how does one prove this?   VAN DER WAERDEN§

Perhaps a more discerning assessment of the situation was expressed by J. R. Newman when he wrote,

A sound appraisal of Egyptian mathematics depends upon a much broader and deeper understanding of human culture than either Egyptologists or historians of science are wont to recognise.‖

* "Mathematics in Ancient Egypt," *Bulletin of the John Rylands Library*, Vol. 15, No. 2 (Manchester, 1931), pp. 439, 440, 441.
† "The Nature of Mathematics," in *The World of Mathematics*, James R. Newman, editor, Vol. 1, Simon and Schuster, New York, 1956, p. 12.
‡ *A Concise History of Mathematics*, Dover, New York, 1948, p. 23.
§ *Science Awakening*, Arnold Dresden, translator, Noordhoff, Groningen, 1954, pp. 31, 36.
‖ *The World of Mathematics*, Vol. 1, p. 178.

Let us see if we can develop and amplify Newman's statement some-what, and perhaps dispel some of the more depressing opinions of some of the less charitable commentators.

It is true that the Egyptians did not show exactly how they estab-lished their rules or formulas, nor how they arrived at their methods in dealing with specific values of the variable. But they nearly always proved that the numerical solution to the problem at hand was indeed correct for the particular value or values they had chosen. For them, this constituted both method and proof, so that many of their solu-tions concluded with sentences like the following:

The producing of the same.   (RMP 4)

The manner of the reckoning of it.   (RMP 41)

The correct procedure for this [type of] problem.   (MMP 9)

Manner of working out.   (RMP 43, 44, 46)

Behold! Does one according to the like for every uneven fraction which may occur.   (RMP 61B)

Thus findest thou the area.   (RMP 55)

These are the correct and proper proceedings.   (MMP 6)

The doing as it occurs. [or] That is how you do it.   (RMP 28 and 23 other problems)

Shalt do thou according to the like in relation to what is said to thee, all like example this.   (RMP 66)

Twentieth-century students of the history and philosophy of science, in considering the contributions of the ancient Egyptians, incline to the modern attitude that an argument or logical proof must be *symbolic* if it is to be regarded as rigorous, and that one or two specific examples using selected numbers cannot claim to be scientifically sound. But this is not true! A nonsymbolic argument or proof *can* be quite rigorous when given for a particular value of the variable; the conditions for rigor are that the particular value of the variable should be *typical*, and that a further generalization to *any* value should be *immediate*. In any of the topics mentioned in this book where the scribes' treatment follows such lines, both these requirements are satisfied, so that the arguments adduced by the scribes are already *rigorous*; the concluding proofs are really not necessary, only confirmatory. The rigor is implicit in the method.

We have to accept the circumstance that the Egyptians did not think and reason as the Greeks did. If they found some exact method (however they may have discovered it), they did not ask themselves *why* it worked. They did not seek to establish its universal truth by an a priori symbolic argument that would show clearly and logically their thought processes. What they did was to explain and define in an ordered sequence the steps necessary in the proper procedure, and at the conclusion they added a verification or proof that the steps outlined did indeed lead to a correct solution of the problem. This was science as they knew it, and it is not proper or fitting that we of the twentieth century should compare too critically their methods with those of the Greeks or any other nation of later emergence, who, as it were, stood on their shoulders. We tend to forget that they were a people who had no plus, minus, multiplication, or division signs, no equals or square-root signs, no zero and no decimal point, no coinage, no indices, and no means of writing even the common fraction p/q; in fact, nothing even approaching a mathematical notation, nothing beyond a very complete knowledge of a twice-times table, and the ability to find two-thirds of any number, whether integral or fractional. With these restrictions they reached a relatively high level of mathematical sophistication.

APPENDIX 2
THE EGYPTIAN CALENDAR

The Egyptian year consisted of 12 months of 30 days each, or 36 decades of 10 days each, with 5 "year-end" (*epagomenal*) days that were dedicated to the gods Osiris, Horus, Seth, Isis, and Nephthys; the year-end days were their gods' birthdays. The 12 months were divided into 3 seasons of 4 months each, which were the *inundation* or sowing period, the *coming-forth* or growing period, and the *summer* or harvest period. This civil year of 365 days was retained because there was no break in its continuity, and was still used in later Hellenistic times; indeed, it was used in the Middle Ages by Copernicus and other astronomers. "This calendar," writes Neugebauer, "is indeed the only intelligent calendar which ever existed in human history."[*] It is simpler even than the "perpetual calendar,"[†] which, though recommended for worldwide use by astronomers, seems condemned to remain forever in some official pigeon-holes in all countries. Now, the true length of the year (the solar year) is 365¼ days,[‡] as the Egyptians knew. This meant that their civil year slipped steadily backwards through the solar year, 1 day every 4 years, 30 days or 1 month every 120 years, and thus 12 months or 1 year every 1,440 years,[§] and the seasons therefore were very slowly changing, though not specially noticeably in one man's lifetime. What did it matter? There were few worries in the ordinary affairs of daily life.

The most important event in Egyptian life was the annual flooding of the Nile River, the inundation period, which coincided pretty closely with the heliacal rising (just before dawn) of Sirius, the Dog Star, the brightest star in either hemisphere. Thus the solar year became also the Sirius year, by which sowing and agriculture generally was controlled. The first brief appearance of Sirius in the eastern sky was an important event in the Egyptian year. The next morning Sirius would appear some minutes earlier, and so on, so that before long, Sirius would no longer herald the dawn, and some other bright star would serve this purpose. This measuring of the days by

* O. Neugebauer, *The Exact Sciences in Antiquity*, Harper, New York, 1962, p. 81.
† "ANZAAS Committee on Calendar Reform," *Australian Journal of Science*, December, 1943; R. J. Gillings, "Perpetual World Calendar," *Australian Mathematics Teacher*, Vol. 1, No. 1, (April, 1945), p. 24.
‡ Less 11 minutes.
§ Approximately of course. The cycle would then start again.

the heliacal rising of stars gave rise to the system of *decans*, in which each chosen star would serve its duty of noting the last hour of night for 10 days (or nights), so that there would be 36 decans distributed through the mornings of the year. Of course, not all decans would be visible through any given night. At the time of the inundation, when Sirius rises heliacally, 12 decans rise during the night, and thus the "hours" of the summer night were determined. In winter there would be more decans visible; thus the length of hours varied slightly, both for the seasons and for nighttime and daytime. We see here the origin of the division of the *day* into 24 hours that is now universally adopted. These two calendars (of 365 and 365¼ days) existing side by side from, it is thought, the time of the first pharaoh of Upper and Lower Egypt, was "the most scientific organisation of calendars which has yet been used by man."*

* J. W. S. Sewell, "The Calendars and Chronology," in *The Legacy of Egypt*, S. R. K. Glanville, editor, Oxford University Press, London, 1963, p. 7.

GREAT PYRAMID MYSTICISM

Perhaps the most famous and best known of all the architectural constructions of the ancient Egyptians are the pyramids, the Great Sphinx, and the Temple of Karnak. And of the 80 or so pyramids, there is no doubt that the Great Pyramid of Khufu (in Greek, Cheops), which was built during the Fourth Dynasty (c. 2644 B.C.), is the one which has most stirred the thoughts and fired the imaginations of all interested in ancient Egypt. Authors, novelists, journalists, and writers of fiction found during the nineteenth century a new topic, a new idea to develop, and the less that was known and clearly understood about the subject, the more freely could they give rein to imagination and invention. These writers were forerunners of the American, Edgar Rice Burroughs, who created the fictional character Tarzan of the Apes, and set him up in central Africa, a country Burroughs had never visited and knew nothing about. Burroughs let his imagination run riot, and his novels (translated into fifty-six languages) achieved sales that were exceeded only by the Bible and Euclid's *Elements*.

Many writers have propounded theories on the origins, the mathematical properties, and the pseudo-astronomical marvels of Cheops, and further, made extravagant prophecies about the Great Pyramid. A resurgence of this cult occurred when Carter discovered the tomb of Tutankhamen in 1923, so that these fictions were presented all over again to the general reading public. Some of them still live on!

It may therefore come as a surprise to those readers with whom any memories remain of those "wonderful disclosures" that most of the miraculous stories written by these writers have no foundation in scientific fact at all; that the remarkable mathematical properties attributed to the Great Pyramid measurements are nowhere attested by scholarly Egyptological studies. It is only because they then were, and perhaps still are, so widely distributed and accepted that any reference to them at all is made in this book. And if some long-cherished illusions are thus destroyed, it is simply because they truly deserve to be destroyed, as being entirely contrary to fact, to history, and to truth.

Among the extraordinary things claimed by these writers, one can read that Piazzi Smyth asserted that half the distance round the square base of the Great Pyramid divided by its height was exactly equal to π, the ratio of the circumference to the diameter of a circle; and that $\frac{1}{360}$ part of the base equals one five-millionth part of the earth's axis of rotation, whatever that might mean!*

Even such a sober-minded person as H. W. Turnbull writes, "Their land surveyors were known as rope stretchers, because they used ropes with knots or marks at equal intervals to measure their plots of land. By this simple means, they were able to construct right angles, for they knew that three ropes of lengths three, four, and five units respectively, could be formed into a right-angled triangle."†

It is, however, nowhere attested that the ancient Egyptians knew even the very simplest case of Pythagoras's theorem! But Turnbull goes further: "As Professor D'Arcy Thompson has suggested, the very shape of the Great Pyramid indicates a considerable familiarity with that [sic] of the regular pentagon. A certain obscure passage in Herodotus, can, by the slightest literal emendation, be made to yield excellent sense. It would imply that the area of each triangular face of the Pyramid, is equal to the square of the vertical height. If this is so, the ratios of height, slope, and base, can be expressed in terms of the *golden section*, or of the ratio of a circle to the side of the inscribed decagon."

I am unable to understand exactly what Turnbull means by this last sentence. But whatever it means, with further slight emendations, the dimensions of the Eiffel Tower or Boulder Dam could be made to produce equally vague and pretentious expressions of a mathematical connotation. I am also unable to locate the reference attributed to D'Arcy Thompson. I would say it is certainly not in his well-known "Growth and Form."‡ Nor can I locate the "certain obscure passage

* Piazzi Smyth was Astronomer Royal of Scotland in the late nineteenth century. His discussion of the Great Pyramid may be found in his *Our Inheritance in the Great Pyramid*, London, 1877.
† H. W. Turnbull, *The Great Mathematicians*, 4th edition, Methuen, London, 1951, pp. 2f.
‡ Unless it be the brief reference on p. 931 of Vol. 2. There, however, Thompson makes no such suggestion: "The *sectio aurea* or 'golden mean' of unity, $(\sqrt{5} - 1)/2 = 0.61803 \ldots$, is a number beloved of the circle squarer, and of all of those who seek to find, and then to penetrate, the secrets of the Great Pyramid. It is deep set in the regular pentagon and

in Herodotus," to find out what this slight "literal emendation" might be.

Anyone wishing to look further into this pyramid mysticism should refer to Leonard Cottrell's *The Mountains of Pharaoh*,* whose Chapter 11 is titled, "The Great Pyramidiot," meaning Smyth but which refers also to John Taylor, Rev. John Davidson, and Edgar Stewart. For a readily available text, *The Pyramids of Egypt*, by I. E. S. Edwards† of the British Museum, can be consulted. That doyen of bibliographers, Raymond Clare Archibald, records the following in his bibliography to *The Rhind Mathematical Papyrus*, which are "enthusiastic in support of the pyramid mysticism of Taylor and Piazzi Smyth":

M. Eyth
Der Kampf um die Cheopspyramide, Heidelberg, 1902.

O. Nairtz
"Die Cheopspyramide, ein viertausendjahriges Räthsel," *Prometheus*, Vol. 17 (1906).

H. Neikes
Der goldene Schnitt und die Geheimnisse der Cheops-Pyramide, Cologne, 1907.

J. and M. Edgar
The Great Pyramid Passages and Chambers, Glasgow, 1910.

K. Kleppisch
Die Cheopspyramide, ein Denkmal mathematischer Erkenntnis, Munich and Berlin, 1921.

F. Noetling
Die kosmischen Zahlen der Cheopspyramide der mathematische Schlüssel zu den Einheits-Gesetzen im Aufbau des Weltalls, Stuttgart, 1921.

D. Davidson and H. Aldersmith
The Great Pyramid, its Divine Message, London, 1924.

---

dodecahedron, the triumphs of Pythagorean or Euclidean geometry. It is a number which becomes by the addition of unity its own reciprocal—its properties never end." *On Growth and Form*, Cambridge University Press, London, 1951.
* Pan Books, London, 1963.
† Penguin Books, London, 1952.

REGARDING MORRIS KLINE'S VIEWS IN
MATHEMATICS, A CULTURAL APPROACH

Morris Kline, in his book on the history of mathematics,* has produced a scholarly volume of seven hundred pages. He is professor of mathematics at New York University. In his opening chapters he speaks of the mathematics of the early civilizations of Egypt, Babylon, India, and China, the last two of which he dismisses very briefly, but then he devotes less than three pages to Egypt and Babylonia. Three pages, to cover an expanse of three thousand years!

Of course, Professor Kline might think that from a "cultural" point of view, these civilizations merited no more than three pages in seven hundred. But concerning Egypt, he mentions briefly the Egyptian equivalent of π, the areas of fields, the volumes of structures, the quantity of material needed to erect pyramids, the grain required to make beer of a certain alcoholic content, Herodotus's remarks about the flooding of the Nile, geometry, and the Egyptian calendar. He then praises their achievements compared with contemporary civilizations, noting that they reached relatively high levels in religion, art, architecture, metallurgy, chemistry, and astronomy, but considers that their "contributions to mathematics were almost insignificant." This pretty sweeping summing up of the Egyptian culture is followed by the astounding statement that, compared with the Greeks, "The mathematics of the Egyptians and Babylonians, is the scrawling of children just learning how to write, as opposed to great literature"! He claims that, "they barely recognized mathematics as a distinct subject" (he apparently has not heard of the Rhind Mathematical Papyrus) and that, "over a period of 4,000 years hardly any progress was made in the subject."†

Such aggravated and condemnatory statements by an author of the stature of Professor Kline can only imply that he has not fully informed himself of the extent and nature of either Egyptian or Babylonian mathematics. And this is most surprising, for he knew at least of Neugebauer and Eves, because he lists their books‡ in his work.

* Morris Kline, *Mathematics, A Cultural Approach*, Addison-Wesley, Reading, Mass., 1962.
† All quotes cited here, *Ibid.*, p. 14.
‡ O. Neugebauer, *The Exact Sciences in Antiquity*, Princeton University Press, Princeton, 1952; H. Eves, *An Introduction to the History of Mathematics*, Rhinehart, New York, 1953.

The *Exact Sciences in Antiquity* alone would have directed his attention to the *Rhind Mathematical Papyrus* of Chace et al. in two large
volumes, published conveniently in Ohio more than a quarter of a
century earlier; the Egyptian mathematics discussed therein most
certainly would not have come under the heading of "the scrawlings
of a child just learning how to write." If then Professor Kline were not
familiar with, or was unaware of, the writings of Eisenlohr, Peet,
Chace, Struve, Griffith, Schack-Schackenburg, Van der Waerden,
Vogel, and others on Egyptian mathematics, then he would have been
wiser to omit any reference to this ancient civilization in his "cultural
approach," and to have devoted his opening chapters to the early
Greeks, about whose work he was certainly very well informed.

APPENDIX 5
THE PYTHAGOREAN THEOREM
IN ANCIENT EGYPT

There is no document to prove that the Egyptian knew even a particular case of the Pythagorean theorem. R. C. ARCHIBALD*

In 90% of all the books, one finds the statement that the Egyptians knew the right triangle of sides 3, 4 and 5, and that they used it for laying out right angles. How much value has this statement? None! B. L. VAN DER WAERDEN†

There is no indication that the Egyptians had any notion even of the Pythagorean theorem, despite some unfounded stories about "harpedonaptai," who supposedly constructed right triangles with the aid of a string with 3 + 4 + 5 = 12 knots [*sic*]. DIRK J. STRUIK‡

The historian Cantor had conjectured that the Egyptians knew that a (3, 4, 5)-triangle is right-angled, and that they used this knowledge to construct right angles. A. SEIDENBERG§

There seems to be no evidence that they knew that triangle (3, 4, 5) is *right-angled*; indeed, according to the latest authority (T. Eric Peet, *The Rhind Mathematical Papyrus*, 1923), nothing in Egyptian mathematics suggests that the Egyptians were acquainted with this or any special cases of the Pythagorean theorem. T. L. HEATH‖

* "Outline of the History of Mathematics," *American Mathematical Monthly*, Vol. 56, No. 1 (January, 1949), p. 16.
† *Science Awakening*, Noordhoff, Groningen, 1954, p. 6.
‡ *Concise History of Mathematics*, Dover, New York, 1948, p. 23.
§ *Scripta Mathematica*, Vol. 24, No. 2 (New York, June, 1959), p. 10, footnote.
‖ *The Thirteen Books of Euclid's Elements*, Vol. 1, Cambridge University Press, London, 1962, p. 352.

APPENDIX 6
THE CONTENTS OF THE RHIND
MATHEMATICAL PAPYRUS

The Recto contains the result of the division of 2 by the 50 odd numbers from 3 to 101. This is followed by a table of the division of the numbers 1 to 9 by 10, expressed in unit fractions.

PROBLEMS 1–6
The division of 1, 2, 6, 7, 8, 9, loaves among 10 men.

PROBLEMS 7–20
The multiplication of (1 $\bar{2}$ $\bar{4}$) and (1 $\bar{\bar{3}}$ $\bar{3}$) by various multipliers containing unit fractions.

PROBLEMS 21–23
Examples in subtraction. 1 − ($\bar{\bar{3}}$ $\overline{15}$), 1 − ($\bar{\bar{3}}$ $\overline{30}$), $\bar{\bar{3}}$ − ($\bar{4}$ $\bar{8}$ $\overline{10}$ $\overline{30}$ $\overline{45}$).

PROBLEMS 24–27
Solution of equations in one unknown of the first degree, resolved by the method of false assumption.

PROBLEMS 28–29
"Think of a Number" problems.

PROBLEMS 30–34
More difficult equations of one unknown of the first degree, resolved by the method of division.

PROBLEMS 35–38
Even more difficult equations in one unknown of the first degree, resolved by the method of false assumption but set in terms of hekats of grain, containers, and hekat measures.

PROBLEMS 39–40
Arithmetic progressions.

PROBLEMS 41–46
Volumes or contents of rectangular and cylindrical granaries.

PROBLEM 47
A table of the fractions of a hekat in Horus-eye fractions.

PROBLEMS 48–55
Areas of triangles, rectangles, trapezia, and circles.

PROBLEMS 56–60

Sekeds, altitudes, and bases of pyramids.

PROBLEMS 61–61B

Tables of and rule for finding two-thirds of odd and even unit fractions.

PROBLEM 62

A rather vague problem in proportion, concerning precious metals by weight.

PROBLEM 63

The proportional division of loaves among men.

PROBLEM 64

An arithmetic progression, and $S_n = (n/2)[2l - (n - 1)d]$.

PROBLEM 65

The proportional division of loaves among men.

PROBLEM 66

Division of fat. The amount issued per day.

PROBLEM 67

The proportion of cattle due as tribute.

PROBLEM 68

The proportional division of grain between gangs of men.

PROBLEMS 69–78

The pesus of bread and beer. Exchanges. Inverse proportion. The concept of a harmonic mean.

PROBLEM 79

A geometric progression whose common ratio is 7.

PROBLEMS 80–81

Tables of Horus-eye fractions of grain in terms of hinu.

PROBLEMS 82–84

Unclear problems, dealing with the amounts of food for various domestic animals, as geese, other birds, and oxen.

PROBLEM 85
Enigmatic writing. Upside down on the papyrus.

PROBLEMS 86–87
Memoranda of certain accounts and incidents, not altogether clear, parts of which are missing.

THE CONTENTS OF THE MOSCOW
MATHEMATICAL PAPYRUS

The Moscow Mathematical Papyrus was purchased in 1893 by V. S. Golenishchev from Abd-el-Rasoul, one of the brothers who found the king's mummy at Deir el-Bahri. The papyrus was originally named after its owner, but in 1912 it passed to the Museum of Fine Arts, "*avec toute ma collection, à Moscou, contre une rente viagère, que le Gouvernement Russe s'était engagé de me payer, ma vie durant.*"* After the Russian revolution of 1917, this life annuity was not paid. Golenishchev died in 1947. The papyrus is over 5 meters long and 8 centimeters high, containing 25 problems, many of which are not clearly readable. Turajeff and Tsinserling wrote on MMP 14 in 1917, but in 1930 W. W. Struve published a complete translation and commentary in *Quellen und Studien*, Section A, Quellen, Berlin. The scribe of the MMP in the view of Egyptologists was a very bad writer.

No. Detail.

1	Damaged and unreadable.
2	Damaged and unreadable.
3	A cedar mast. $\overline{3}$ $\overline{5}$ of 30 = 16. Unclear.
4	Area of a triangle. $\overline{2}$ of 4 × 10 = 20.
5	Pesus of loaves and bread. Same as No. 8. See Chapter 12.
6	Rectangle, area = 12, $b = \overline{2}$ $\overline{4}l$. Find l and b.
7	Triangle, area = 20, $h = 2$ $\overline{2}b$. Find h and b.
8	Pesus of loaves and bread. See Chapter 12.
9	Pesus of loaves and bread. See Chapter 12.
10	Area of curved surface of a hemisphere (or cylinder). See Chapter 18.
11	Loaves and basket. Unclear.
12	Pesu of beer. Unclear.
13	Pesus of loaves and beer. Same as No. 9. See Chapter 12.
14	Volume of a truncated pyramid. $V = (h/3)(a^2 + ab + b^2)$. See Chapter 18.
15	Pesu of beer. See Chapter 12.
16	Pesu of beer. Similar to No. 15.
17	Triangle, area = 20, $b = (\overline{3}\ \overline{15})h$. Find h and b.

* "With my whole collection in Moscow, in return for a life annuity which the Russian government promised to pay me as long as I live."

18 Measuring cloth in cubits and palms. Unclear.
19 Solve the equation, $1\ \bar{2}x + 4 = 10$. Clear. See Chapter 14.
20 Pesu of 1,000 loaves. Horus-eye fractions. See Chapter 12.
21 Mixing of sacrificial bread. See Chapter 12.
22 Pesus of loaves and beer. Exchange. See Chapter 12.
23 Computing the work of a cobbler. Unclear. Peet says very difficult.
24 Exchange of loaves and beer. See Chapter 12.
25 Solve the equation, $2x + x = 9$. Elementary and clear.

On analysis of these problems we find,

[2 problems] Nos. 1 and 2 are not readable.

[11] Five problems (8, 9, 13, 22, 24) on the pesus of loaves and beer are not perfectly clear. Three problems (5, 20, 21) deal with the pesu of loaves only. They are difficult to understand. Three problems (12, 15, 16) deal with beer and its pesu only. They are clear and simple.

[6] Three treat the area of a triangle. No. 4 merely finds the area of a right triangle, while Nos. 7 and 17 are equivalent to the solution of two simultaneous equations, one of the second degree. Two problems (19, 25) concern the solution of equations of the first degree, which are very simple, and No. 6 is on simultaneous equations, one of the second degree.

[4] Problems 3, 11, 18, and 23 are miscellaneous problems, none of which is entirely clear.

[2] No. 14 on the volume of a truncated pyramid is a most important problem in the history of Egyptian mathematics. It has no counterpart in any other mathematical papyrus. No. 10 deals, I consider, with the area of the surface of a hemisphere, as Struve thought, and if this is so, it becomes the outstanding Egyptian achievement in the field of mathematics.

A PAPYRITIC MEMO PAD

I have often made references to the probable use of papyritic memo pads by the scribes when we have been confronted with what appears to be a piece of mental arithmetic that seems to be rather too long or too difficult to be done in the head. If such memo pads were sometimes used, we can envisage what one might have looked like with the following example, taken from Problem 35 of the RMP.

The scribe writes at once,

$$\bar{\bar{3}} \text{ of } (\bar{5} \quad \overline{10}) = \bar{5}.$$

How might he have done this on his memo pad?

Either,

$$
\begin{aligned}
\bar{\bar{3}} \text{ of } (\bar{5} \quad \overline{10}) &= (\overline{10} \quad \overline{30}) \quad \overline{15} && \text{[RMP 61B.]} \\
&= \overline{10} \quad (\overline{15} \quad \overline{30}) && \text{[Rearranging.]} \\
&= \overline{10} \qquad \overline{10} && \text{[EMLR 24, or G rule.]} \\
&= \qquad \bar{5}. && \text{[gen. (1, 1).]}
\end{aligned}
$$

Or,

$$
\begin{aligned}
\bar{\bar{3}} \text{ of } (\bar{5} \quad \overline{10}) &= \bar{\bar{3}} \text{ of } (1 \ \bar{2} \text{ of } \bar{5}) && [\overline{10} = \bar{2} \text{ of } \bar{5}.] \\
&= (\bar{\bar{3}} \text{ of } 1 \ \bar{2}) \text{ of } \bar{5} && \text{[Rearranging.]} \\
&= \qquad 1 \text{ of } \bar{5} && [\bar{\bar{3}} = \text{reciprocal of } 1 \ \bar{2}.] \\
&= \qquad \bar{5}.
\end{aligned}
$$

Or,

$$
\begin{aligned}
\bar{\bar{3}} \text{ of } (\bar{5} \quad \overline{10}) &= \bar{\bar{3}} \text{ of } (\overline{10} \quad \overline{10}) \quad \overline{10} && \text{[gen. (1, 1).]} \\
&= \bar{\bar{3}} \text{ of } \overline{10} \quad \overline{10} \quad \overline{10} && \text{[Removing brackets.]} \\
&= \qquad \overline{10} \quad \overline{10} && \text{[2 parts out of 3.]} \\
&= \qquad \bar{5}. && \text{[gen. (1, 1).]}
\end{aligned}
$$

Or,

$$
\begin{aligned}
\bar{\bar{3}} \text{ of } (\bar{5} \quad \overline{10}) &= 1 \ \bar{3} \text{ of } \overline{10} \qquad \overline{20} && \text{[Mult. and divide by 2.]} \\
&= 4 \times (\overline{30} \quad \overline{60}) && \text{[Mult. and divide by 3.]} \\
&= 4 \times \qquad \overline{20} && \text{[EMLR 23, or G rule.]} \\
&= \qquad \bar{5}.
\end{aligned}
$$

HORUS-EYE FRACTIONS IN TERMS OF HINU: PROBLEMS 80, 81 OF THE RHIND MATHEMATICAL PAPYRUS

PROBLEM 80

This is not really a problem but a reference table, which the scribe states is for the use of the "functionaries" of the granary, in conjunction with their vessels for measuring out the grain. The table is based on the equivalence 10 hinu = 1 hekat:

Hekat	Hinu					
$\bar{1}$	10					
$\bar{2}$	5					
$\bar{4}$	2	$\bar{2}$				
$\bar{8}$	1		$\bar{4}$			
$\overline{16}$		$\bar{2}$		$\bar{8}$		
$\overline{32}$			$\bar{4}$		$\overline{16}$	
$\overline{64}$				$\bar{8}$		$\overline{32}$

PROBLEM 81

This is a more extensive table, the first portion of which is an exact repetition of Problem 80. Then follow 28 entries, giving various fractions of a hekat of grain, expressed in terms of hinu and in Horus-eye fractions and ro (see Table A9.1). There are some duplications, as the first column of line numbers (following Chace) in Table A9.1 shows. I have, of course, altered the scribe's sequence in this table.

TABLE A9.1
Horus-eye fractions in terms of hinu.

Line (Chace)	Fraction of a Hekat	$\overline{2}$	$\overline{4}$	$\overline{8}$	$\overline{16}$	$\overline{32}$	$\overline{64}$	ro.	Equivalent in Hinu
a3, d3.	$\overline{\overline{3}}$	$\overline{2}$		$\overline{8}$		$\overline{32}$		3 $\overline{3}$	6 $\overline{\overline{3}}$
c3.	$\overline{2}$	$\overline{2}$							5
a6, d4.	$\overline{3}$		$\overline{4}$		$\overline{16}$		$\overline{64}$	1 $\overline{\overline{3}}$	3 $\overline{3}$
a7, c4.	$\overline{4}$		$\overline{4}$						2 $\overline{2}$
a8, b1.	$\overline{5}$			$\overline{8}$	$\overline{16}$			4	2
a9.	$\overline{6}$			$\overline{8}$		$\overline{32}$		3 $\overline{3}$	1 $\overline{\overline{3}}$
d5.	$\overline{8}$			$\overline{8}$					1 $\overline{4}$
b2.	$\overline{10}$				$\overline{16}$	$\overline{32}$		2	1
b5.	$\overline{15}$				$\overline{16}$			1 $\overline{3}$	$\overline{\overline{3}}$
d6.	$\overline{16}$				$\overline{16}$				$\overline{2}$ $\overline{8}$
b3.	$\overline{20}$					$\overline{32}$	$\overline{64}$	1	$\overline{2}$
c1.	$\overline{30}$					$\overline{32}$		$\overline{3}$	$\overline{3}$
e1.	$\overline{32}$					$\overline{32}$			$\overline{4}$ $\overline{16}$
b4.	$\overline{40}$						$\overline{64}$	3	$\overline{4}$
c2.	$\overline{60}$						$\overline{64}$	$\overline{3}$	$\overline{6}$
e2.	$\overline{64}$						$\overline{64}$		$\overline{8}$ $\overline{32}$
c5.	$\overline{2}$ $\overline{4}$	$\overline{2}$	$\overline{4}$						7 $\overline{2}$
a4, d1.	$\overline{2}$ $\overline{8}$	$\overline{2}$		$\overline{8}$					6 $\overline{4}$
	$\overline{2}$ $\overline{16}$	$\overline{2}$			$\overline{16}$				5 $\overline{2}$ $\overline{8}$
	$\overline{2}$ $\overline{32}$	$\overline{2}$				$\overline{32}$			5 $\overline{4}$ $\overline{16}$
	$\overline{2}$ $\overline{64}$	$\overline{2}$					$\overline{64}$		5 $\overline{8}$ $\overline{32}$
a5, d2.	$\overline{4}$ $\overline{8}$		$\overline{4}$	$\overline{8}$					3 $\overline{2}$ $\overline{4}$
	$\overline{4}$ $\overline{16}$		$\overline{4}$		$\overline{16}$				3 $\overline{8}$
	$\overline{4}$ $\overline{32}$		$\overline{4}$			$\overline{32}$			2 $\overline{2}$ $\overline{4}$ $\overline{16}$
	$\overline{4}$ $\overline{64}$		$\overline{4}$				$\overline{64}$		2 $\overline{2}$ $\overline{8}$ $\overline{32}$
	$\overline{2}$ $\overline{4}$ $\overline{8}$	$\overline{2}$	$\overline{4}$	$\overline{8}$					8 $\overline{2}$ $\overline{4}$
	$\overline{2}$ $\overline{4}$ $\overline{16}$	$\overline{2}$	$\overline{4}$		$\overline{16}$				8 $\overline{8}$
	$\overline{2}$ $\overline{4}$ $\overline{32}$	$\overline{2}$	$\overline{4}$			$\overline{32}$			7 $\overline{2}$ $\overline{4}$ $\overline{16}$
	$\overline{2}$ $\overline{4}$ $\overline{64}$	$\overline{2}$	$\overline{4}$				$\overline{64}$		7 $\overline{2}$ $\overline{8}$ $\overline{32}$

THE EGYPTIAN EQUIVALENT OF THE
LEAST COMMON DENOMINATOR

TABLE A10.1
Least common denominators in the Rhind Mathematical Papyrus.

RMP No.	Fractions to be Added	Reference No.	The Sum
21.	$\overline{\overline{3}}\ \overline{5}\ \overline{15}\ \overline{15}$	15	1
22.	$\overline{\overline{3}}\ \overline{5}\ \overline{10}\ \overline{30}$	30	1
23.	$\overline{4}\ \overline{8}\ \overline{9}\ \overline{10}\ \overline{30}\ \overline{40}\ \overline{45}$	45	$\overline{\overline{3}}$
32.	$\overline{12}\ \overline{18}\ \overline{24}\ \overline{36}\ \overline{48}\ \overline{114}\ \overline{228}$ $\overline{342}\ \overline{456}\ \overline{684}\ \overline{912}$ $\Big\}$	912	$\overline{4}$
33.	$\overline{56}\ \overline{84}\ \overline{112}\ \overline{392}\ \overline{679}\ \overline{776}\ \overline{1164}$ $\overline{1358}\ \overline{4074}\ \overline{4753}\ \overline{1358}\ \overline{1552}\ \overline{5432}$ $\Big\}$	5432	$\overline{21}$
34.	$\overline{7}\ \overline{14}\ \overline{14}\ \overline{28}\ \overline{28}\ \overline{56}$	56	$\overline{4}\ \overline{8}$
36.	$\overline{12}\ \overline{20}\ \overline{30}\ \overline{53}\ \overline{53}\ \overline{106}\ \overline{106}\ \overline{159}$ $\overline{212}\ \overline{265}\ \overline{318}\ \overline{318}\ \overline{530}\ \overline{636}\ \overline{795}\ \overline{1060}$ $\Big\}$	1060	$\overline{4}$
37.	$\overline{16}\ \overline{32}\ \overline{64}\ \overline{72}\ \overline{576}$	576	$\overline{8}$
37.	$\overline{12}\ \overline{16}\ \overline{32}\ \overline{36}\ \overline{36}\ \overline{96}\ \overline{288}\ \overline{288}$	288	$\overline{4}$
38.	$\overline{11}\ \overline{11}\ \overline{22}\ \overline{22}\ \overline{33}\ \overline{66}\ \overline{66}$	66	$\overline{\overline{3}}$

In the group of problems RMP 7 to 20, which show how to multiply certain fractional numbers, there are a further 8 examples of the addition of fractions by the use of red auxiliaries, but the reference numbers are confined to the numbers 18 and 28. But these are a special group. In all of the 10 other examples of the RMP where this method is used, the scribe chose the largest denominator as a reference number, and in four of these (34, 37, 37, 38), it happens to be what we call the least common denominator, but this is just by chance (Table A10.1). The same applies to numbers 21 and 22, but for some reason the auxiliary numbers are in black like all other numbers, not in red; this is probably an accident—the scribe may temporarily have run out of red ink. Nowhere does the scribe show how he did all the lesser divisions, for example, the 16 divisions in RMP 36, where e.g., $1060 \div 159 = 6\ \overline{\overline{3}}$, or $1060 \div 12 = 88\ \overline{\overline{3}}$. Now, however adept the scribe may have been at mental arithmetic, he surely must have made use of memo pads, some kind of papyrus scribbling blocks, to perform these divisions, and to have access to $\overline{\overline{3}}$ tables, for these two divisions would have to be done as follows.

$1060 \div 159$:

	1		159
	\2		\318
	\4		\636
	\$\bar{\bar{3}}$		\106
Totals	6 $\bar{\bar{3}}$		1060.

$1060 \div 12$:

	1		12
	2		24
	4		48
	\8		\ 96
	\16		\192
	32		384
	\64		\768
	$\bar{\bar{3}}$		8
	\$\bar{3}$		\ 4
Totals	88 $\bar{\bar{3}}$		1060.

It is difficult to imagine such divisions as these being done in the head.

To sum up, we find that there are in the RMP about 130 occasions where the scribe needed to add up unit fractions, and quite often a large number of fractions, as in RMP 70, where there are 20 of them,* yet he used his technique of a reference number and the red auxiliaries only 18 times, i.e., in less than 15 percent of the cases where they would be applicable. It seems pretty clear that in Problems 7 to 20, where he was teaching the multiplication of fractions, he included this method 8 times as instruction to the student in very simple cases, as for example (RMP 8), he used the reference number 18 to prove that $\bar{4}\ \bar{6}\ \overline{12} = \bar{2}$. In Problems 21 to 23, where he showed how to subtract fractions, the additions required are a little harder (see table) and he used it 3 times; while in Problems 30 to 34, where he showed how to divide by fractions, he again used it 3 times, one of which was unnecessary. Finally, in Problems 35 to 38 he had occasion to use it 4 times, and possibly a couple of these were unnecessary from

* The last three are $\overline{336}$, $\overline{504}$, and $\overline{1008}$.

the point of view of difficulty. In this last group of problems the scribe was dealing more particularly with the divisions of the hekat for grain, in which Horus-eye fractions were used. I mean by "unnecessary" that in Problem 34, the addition of the fractions

$$\bar{7} \quad \overline{14} \quad \overline{14} \quad \overline{28} \quad \overline{28} \quad \overline{56} = \bar{4} \quad \bar{8}$$

was purely mental for the scribe, because he had just established that

$$\bar{7} \quad \overline{14} \quad \overline{28} = \bar{4} \qquad \text{(Problem 11)}$$
$$\text{and } \overline{14} \quad \overline{28} \quad \overline{56} = \bar{8}, \qquad \text{(Problem 12)}$$

so that the sum followed at once by inspection. But he was teaching a certain technique, so he used his reference number method for the benefit of his readers.

A TABLE OF TWO-TERM EQUALITIES FOR EGYPTIAN UNIT FRACTIONS

TABLE A11.1
Two-term equalities for Egyptian unit fractions.

(1, 1)	2 2 = 1	(2, 1)	6 3 = 2	(3, 1)	12 4 = 3	(4, 1)	20 5 = 4	(5, 1)	30 6 = 5
(1, 2)	3 6 = 2	(2, 2)	8 8 = 4	(3, 2)	15 10 = 6	(4, 2)	24 12 = 8	(5, 2)	35 14 = 10
(1, 3)	4 12 = 3	(2, 3)	10 15 = 6	(3, 3)	18 18 = 9	(4, 3)	28 21 = 12	(5, 3)	40 24 = 15
(1, 4)	5 20 = 4	(2, 4)	12 24 = 8	(3, 4)	21 28 = 12	(4, 4)	32 32 = 16	(5, 4)	45 36 = 20
(1, 5)	6 30 = 5	(2, 5)	14 35 = 10	(3, 5)	24 40 = 15	(4, 5)	36 45 = 20	(5, 5)	50 50 = 25
(1, 6)	7 42 = 6	(2, 6)	16 48 = 12	(3, 6)	27 54 = 18	(4, 6)	40 60 = 24	(5, 6)	55 66 = 30
(1, 7)	8 56 = 7	(2, 7)	18 63 = 14	(3, 7)	30 70 = 21	(4, 7)	44 77 = 28	(5, 7)	60 84 = 35
(1, 8)	9 72 = 8	(2, 8)	20 80 = 16	(3, 8)	33 88 = 24	(4, 8)	48 96 = 32	(5, 8)	65 104 = 40
(1, 9)	10 90 = 9	(2, 9)	22 99 = 18	(3, 9)	36 108 = 27	(4, 9)	52 117 = 36	(5, 9)	70 126 = 45
(1, 10)	11 110 = 10	(2, 10)	24 120 = 20	(3, 10)	39 130 = 30	(4, 10)	56 140 = 40	(5, 10)	75 150 = 50
(6, 1)	42 7 = 6	(7, 1)	56 8 = 7	(8, 1)	72 9 = 8	(9, 1)	90 10 = 9	(10, 1)	110 11 = 10
(6, 2)	48 16 = 12	(7, 2)	63 18 = 14	(8, 2)	80 20 = 16	(9, 2)	99 22 = 18	(10, 2)	120 24 = 20
(6, 3)	54 27 = 18	(7, 3)	70 30 = 21	(8, 3)	88 33 = 24	(9, 3)	108 36 = 27	(10, 3)	130 39 = 30
(6, 4)	60 40 = 24	(7, 4)	77 44 = 28	(8, 4)	96 48 = 32	(9, 4)	117 52 = 36	(10, 4)	140 56 = 40
(6, 5)	66 55 = 30	(7, 5)	84 60 = 35	(8, 5)	104 65 = 40	(9, 5)	126 70 = 45	(10, 5)	150 75 = 50
(6, 6)	72 72 = 36	(7, 6)	91 78 = 42	(8, 6)	112 84 = 48	(9, 6)	135 90 = 54	(10, 6)	160 96 = 60
(6, 7)	78 91 = 42	(7, 7)	98 98 = 49	(8, 7)	120 105 = 56	(9, 7)	144 112 = 63	(10, 7)	170 119 = 70
(6, 8)	84 112 = 48	(7, 8)	105 120 = 56	(8, 8)	128 128 = 64	(9, 8)	153 136 = 72	(10, 8)	180 144 = 80
(6, 9)	90 135 = 54	(7, 9)	112 144 = 63	(8, 9)	136 153 = 72	(9, 9)	162 162 = 81	(10, 9)	190 171 = 90
(6, 10)	96 160 = 60	(7, 10)	119 170 = 70	(8, 10)	144 180 = 80	(9, 10)	171 190 = 90	(10, 10)	200 200 = 100

APPENDIX 12
TABLES OF HIERATIC INTEGERS AND FRACTIONS, SHOWING VARIATIONS

TABLE A12.1
Integers as written in various papyri.

		Reisner	MMP	KP	EMLR	RMP	Berlin
I	1						
II	2						
III	3						
IIII	4						
(5)	5						
(6)	6						
(7)	7						
(8)	8						
(9)	9						
∩	10						
∩∩	20						
∩∩∩	30						
∩∩∩∩	40						
50	50						
60	60						
70	70						
80	80						
90	90						

TABLE A12.1 (*continued*)

		Reisner	MMP	KP	EMLR	RMP	Berlin
୬	100						
୬୬	200						
୬୬୬	300						
୬୬	400						
୬୬୬	500						
୬୬୬	600						
୬୬୬୬	700						
୬୬୬୬	800						
୬୬୬	900						
	1000						
	2000						
	3000						
	4000						

9.

TABLE A12.2
Hieratic integers, showing variations.

1	31	61	91	
2	32	62	92	
3	33	63	93	
4	34	64	94	
5	35	65	95	
6	36	66	96	
7	37	67	97	
8	38	68	98	
9	39	69	99	
10	40	70	100	
11	41	71	200	
12	42	72	300	
13	43	73	400	
14	44	74	500	
15	45	75	600	
16	46	76	700	
17	47	77	800	
18	48	78	900	
19	49	79	1000	
20	50	80	2000	

TABLE A12.2 (*continued*)

21	51	81	3,000	
22	52	82	4,000	
23	53	83	5,000	
24	54	84	6,000	
25	55	85	8,000	
26	56	86	9,000	
27	57	87	10,000	
28	58	88	100,000	
29	59	89	200,000 / 300,000	
30	60	90	400,000 / 500,000	

TABLE A12.3
Hieratic fractions, showing variations.

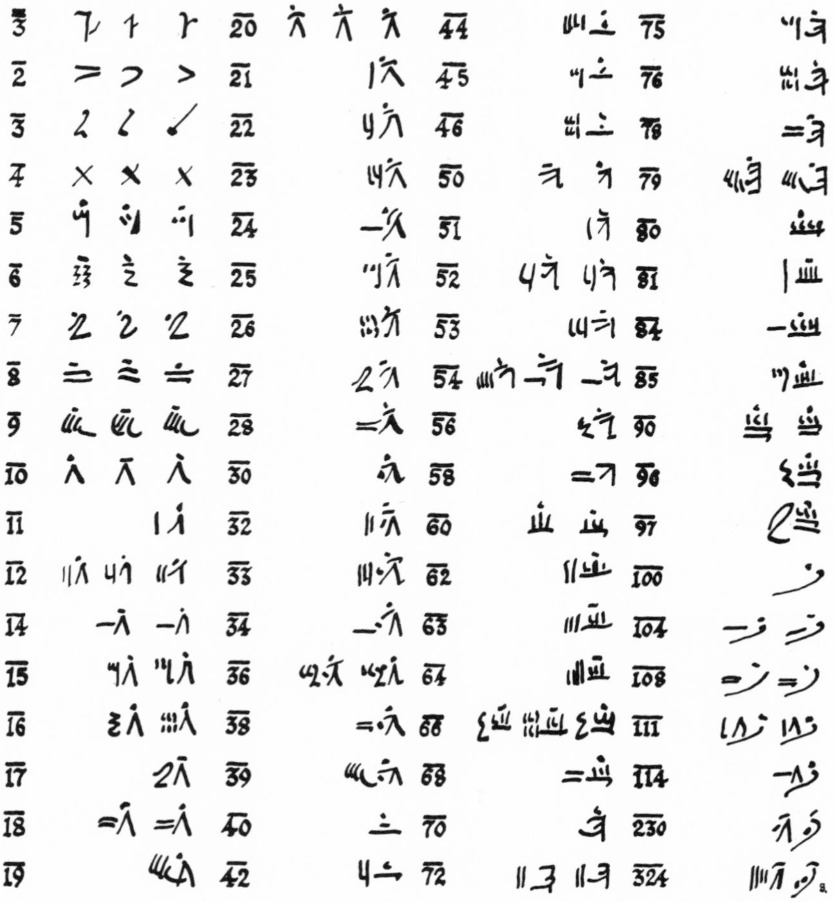

frac	glyphs	frac	glyphs	frac	glyphs	frac	glyphs
$\overline{3}$	⅂ ⅂ ⅂	$\overline{20}$	λ λ λ	$\overline{44}$	�III ⹁	$\overline{75}$	⹁⹁
$\overline{2}$	⹀ ⹁ >	$\overline{21}$	⎮λ	$\overline{45}$	⹁⹀	$\overline{76}$	⹁⹁
$\overline{3}$	⹁ ⹁ ⹁	$\overline{22}$	⹁λ	$\overline{46}$	⹁λ	$\overline{78}$	⹀⹁
$\overline{4}$	X X X	$\overline{23}$	⹁λ	$\overline{50}$	⹁ ⹁	$\overline{79}$	⹁⹁ ⹁⹁
$\overline{5}$	⹁ ⹁ ⹁	$\overline{24}$	⹁λ	$\overline{51}$	⹁λ	$\overline{80}$	⹁⹁
$\overline{6}$	⹁ ⹁ ⹁	$\overline{25}$	⹁λ	$\overline{52}$	⹁λ	$\overline{81}$	⹁⹁ ⹁⹁
$\overline{7}$	⹁ ⹁ ⹁	$\overline{26}$	⹁λ	$\overline{53}$	⹁λ	$\overline{84}$	⹁⹁
$\overline{8}$	⹀ ⹁ ⹀	$\overline{27}$	⹁λ	$\overline{54}$	⹁λ	$\overline{85}$	⹁⹁ ⹁⹁ ⹁⹁
$\overline{9}$	⹁ ⹁ ⹁	$\overline{28}$	⹀λ	$\overline{56}$	⹁λ	$\overline{90}$	⹁⹁
$\overline{10}$	λ λ λ	$\overline{30}$	⹁λ	$\overline{58}$	⹁λ	$\overline{96}$	⹀⹁
$\overline{11}$	⎮λ	$\overline{32}$	⎮⎮λ	$\overline{60}$	⎮⎮λ	$\overline{97}$	⹁⹁ ⹁⹁
$\overline{12}$	⎮⎮λ ⹁⹁ ⎮⎮⹁	$\overline{33}$	⎮⎮⎮λ	$\overline{62}$	⎮⎮⎮λ	$\overline{100}$	⎮⎮⹁
$\overline{14}$	⹀λ ⹀λ	$\overline{34}$	⹁λ	$\overline{63}$	⹁λ	$\overline{104}$	⎮⎮⎮⹁
$\overline{15}$	⹁λ ⹁λ	$\overline{36}$	⹁λ	$\overline{64}$	⹁λ ⹁λ	$\overline{108}$	⎮⎮⎮⎮⹁
$\overline{16}$	⹁λ ⹁λ	$\overline{38}$	⹁λ	$\overline{66}$	⹁λ	$\overline{111}$	⹁⹁ ⹁⹁ ⹁⹁ ⹁⹁
$\overline{17}$	⹁λ	$\overline{39}$	⹁λ	$\overline{68}$	⹁λ	$\overline{114}$	⹀⹁
$\overline{18}$	⹀λ ⹀λ	$\overline{40}$	⹀λ	$\overline{70}$	⹀	$\overline{230}$	⹁⹁
$\overline{19}$	⹁λ	$\overline{42}$	⹁λ	$\overline{72}$	⹁⹀	$\overline{324}$	⎮⎮⹁ ⎮⎮⹁

TABLE A13.1 Chronology.

Dynasties	Approx. no. of pharaohs	Approx. aggregate totals	Neugebauer—Parker	Gardiner	Edwards	Cottrell	Seidl	Purnell	Breasted	Approx. duration	Approx. aggregate totals	Estimated occurrence within a century
Archaic	6	6							4241	?	?	Calendar. Metal cast.
Early Dynastic			B.C.									
I	8	14	3110	3100 ±150	3188	3100	3188	3200	3400	250	250	Menes—Upper & Lower Egypt.
II	9	23	2883	2847		2780				150	400	Capital—Memphis. Zoser—Step pyramid—Sakkara.
Old Kingdom												
III	9	32	2664	2700	2815	2670	2815	2778	2980	80	480	Copper tools. Hieroglyphic—Hieratic writing.
IV	8	40	2614	2620		2600	2690	2680	2900	140	620	Cheops—Great Pyramid—Giza. Cephren Pyramid—Sphinx—Giza. Mycerinus Pyramid—Giza.
V	8	48	2501	2480		2560		2565	2750	140	760	Neferikara Pyramid—Abu Sir.
VI	11	59	2341	2340		2420	2420	2423	2625	160	920	Nerenra Pyramid—Sakkara. Pepi II Pyramid—Sakkara. World's longest reigning king. 90+ yrs.
First Intermediate Period												
VII	?	?	?	?	2294	2270	2294	2263	2475	—	—	

Dynasty			B.C. dates (various authorities)						Years	Total	Notes
VIII	18	77	2174						65	985	
IX	14	91	2154			2445			73	1058	
X	18	109	2100		*Interregnum* 2239 / 2233				68	1126	
Middle Kingdom											
XI	8	117	2134	2134	2133	2134	2160		143	1269	Capital—Thebes. Reisner Papyri—Hieratic—Boston. Akhmim tablets—Hieratic—Cairo. Senusret I Pyramid—Lisht.
XII	8	125	1991	1991	2000	1990	1891	2000	205	1474	Rhind Mathematical Papyrus—Hieratic—British Museum. Moscow Mathematical Papyrus—Hieratic—Moscow.
Second Intermediate Period											
XIII			1785	1786	1777		1788	1777			
Hyksos (Shepherd Kings) Period											
XIV		132	1785	1700	1700						Kahun Papyri—Hieratic—London. Hyksos Domination.
XV		257	1678								
XVI			1647	1680					211	1685	Rhind Mathematical Papyrus copied—Aᶜḥ-mosè. Egyptian Mathematical Leather Roll—British Museum.
XVII			1600	1600	1570						Hyksos expelled.

TABLE A13.1 (*continued*)

Dynasties	Approx. no. of pharaohs	Approx. aggregate totals	Neugebauer—Parker	Gardiner	Edwards	Cottrell	Seidl	Purnell	Breasted	Approx. duration	Approx. aggregate totals	Estimated occurrence within a century
New Kingdom												
XVIII	16	273	1570	1575	1573	1567	1573	1580	1580	267	1952	Q. Hatshepsut. Bronze appears. Tutmosis. Akhenaten—Nefertiti. Tutankhamen. Ramses II. The Exodus.
XIX	9	282	1304	1308		1320	1314	1320	1350	124	2076	Berlin Papyri—Hieratic—Berlin. Anastasy Papyrus—Hieratic—Leyden.
XX	10	292	1192	1184		1200	1194	1250	1200	100	2176	Temple of Karnak—Abu Simbel.
Late Dynastic Period												
XXI	7	299	1075	1087	1090	1085		1085	1090	142	2318	Tanites. Religious writing still Hieroglyphic. Other, Hieratic.
XXII	8	307	940	945				950	945	128	2446	Libyans.
XXIII	4	311	761	817		745			745	87	2535	
XXIV	{2	313	725	730		718			718	5	2538 ⎫	Overlapping period.
XXV	{6	319	736	725		712	712	751	712	61	2599 ⎭	Ethiopian conquest. Demotic writing.
XXVI	9	328	664	664	663	664		656	663	140	2739	Saite period. Assyrian domination.

Period / Dynasty											Events
Persian Conquest XXVII	8	336	525	525	525	525	525	525	120	2859	Persian conquest. Cambyses—Xerxes—Darius. Thales—Pythagoras.
Late Dynastic Period											
XXVIII	1	337	404	404	404	404	404		5	2864	Herodotus visits Egypt (450).
XXIX	5	342	398	399	398	398	399	398	20	2884	
XXX	3	345	378	380	378	380		38		2922	
XXXI	3	348	341	343	332	332	332	332	10	2932	Aristotle. Three Persian Kings.
Greek Period											
Alexander the Ptolemies {	14		332	332	332	332	332	332	9	2941	Capital Alexandria—Coptic writing. Archimedes.
		362	323	323	323	323	323		293	3234	Cleopatra—Last of the Ptolemies (30).
Roman Period											
Province of			30	30	30	30	30	30	30	3264	Julius Caesar—Mark Antony. Brass appears.
Rome {		0	0	0	0	0	0	0		3294	Birth of Christ. Revillout Papyrus—Demotic—London (140).
		324 A.D.	364	364	364	364	364	364	A.D. 400	3658	Thompson Tablet—Greek—London (350).
										3694	
								500		3794	Michigan Mathematical Papyrus—Greek—Ann Arbor (350).

TABLE A13.1. (continued)

Dynasties	Approx. no. of pharaohs	Approx. aggregate totals	Neugebauer—Parker	Gardiner	Edwards	Cottrell	Seidl	Purnell	Breasted	Approx. duration	Approx. aggregate totals	Estimated occurrence within a century
Byzantine Empire												
Byzantine Control			640						600		3894	Justinian (538). Crum Ostracon—Coptic—London (550).
									700		3994	Hegira (622)—Mohammed—Flight from Mecca. Alexandria captured by the Arabs (642).
Islam									800		4094	Baillet-Akhmim Papyrus—Greek—Cairo (750).
									900		4194	Crum.—Parchment—Palimpsest. Demotic & Greek—
									1000		4294	London (circa 950–1000).
									1100		4394	
									1200		4494	Saladin—Sultan of Egypt (1137–1193).
Mame-lukes									1300		4594	Mamelukes conquer Egypt 1250.
									1400		4694	End of Byzantine Empire (1453). Constantinople captured by Ottoman Turks.
									1500		4794	Mamelukes defeated (1517).

1600	4894	Robert Recorde—Arithmetic. Galileo. Napier—Logarithms. Descartes—Indices. Pepys (1662) learns multiplication tables (Age 30).
1700	4994	Newton—Principia (1687). Calculus. Leibnitz. Napoleon's Egyptian expedition (1789).
1800	5094	Rosetta stone found—Nile delta (1799). Young (1814).—Bankes—Philae Obelisk (1819). Champollion—"Je tiens l'affaire." (1822).
1900	5194	RMP found—Thebes (1859).—Suez Canal (1869). Flinders Petrie. Tutankhamen's tomb—Carter (1922).
2000	5294	Chace—RMP (1927). K. Fuad (1936). K. Farouk. Col. Nasser.

Turkish Domination

(Mahomet Ali)

World War I

World War II

362 Pharaohs

62 English Kings

APPENDIX 14
A MAP OF EGYPT

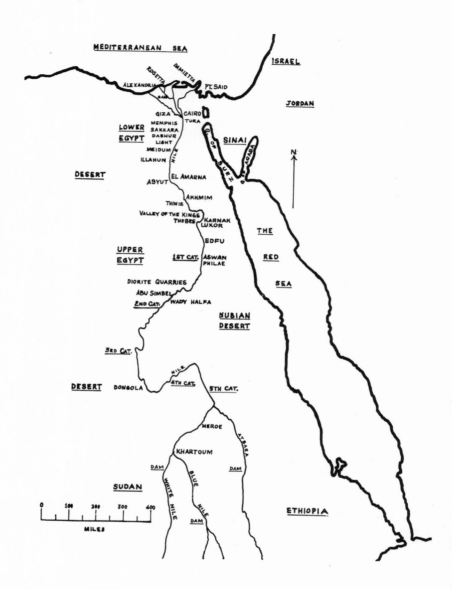

BIBLIOGRAPHY

Akhmîm Papyrus (AP)
The Cairo Museum, Egypt, Catalogue No. 10758.

Archibald, R. C.
Bibliography of Egyptian Mathematics, Mathematical Association of America, Oberlin, Ohio, 1927.

Archibald, R. C.
Chace Memorial Address, The Review Club of Pembroke College, Providence, Rhode Island, 1932.

Archibald, R. C.
"Outline of the History of Mathematics," *The American Mathematical Monthly*, Vol. 56 (Menasha, Wisconsin, 1949).

Becker and Hofmann
Geschichte der Mathematik, Athenäum-Verlag, Bonn, 1951.

Bergamini, David
LIFE *Science Library: Mathematics*, TIME-LIFE International, Netherlands N.V., 1965.

Berlin Papyrus (BP)
Staatliche Museen zu Berlin, Catalogue No. 6619.

Borchardt, Ludwig
Gegen die Zahlenmystik an der grossen Pyramide bei Gise, Berlin, 1922.

Borchardt, Ludwig
"Salary Distribution for the Personnel of the Temple of Illahun," *Zeitschrift für Ägyptische Sprache*, Vol. 40 (Leipzig, 1902–1903), pp. 113–117.

Boyer, Carl B.
A History of Mathematics, John Wiley, New York, 1968.

Breasted, J. H.
A History of Egypt, Hodder and Stoughton, London, 1946.

Broken Hill Proprietary Company
Shapes and Sections Handbook, Melbourne, 1937.

Brugsch, H.
"Über Bau und Masse des Temples von Edfu," *Zeitschrift für Ägyptische Sprache*, Vol. 8 (Leipzig, 1870).

Bruins, E. M.
"Ancient Egyptian Arithmetic," *Kon. Nederland Akademie van Wetenschappen*, Ser. A, Vol. 55, No. 2 (Amsterdam, 1952).

Bruins, E. M.
"Over de Benandering van $\pi/4$ in de Aegyptische Meetkunde," *Kon. Nederland Akademie van Wetenschappen*, Vol. 48 (Amsterdam, 1945).

Budge, Sir E. A. W.
The Rosetta Stone, Trustees of the British Museum, London, 1913. Reprinted with revision, 1957.

Cantor, M.
Vorlesungen über Geschichte der Mathematik, Vol. 1, Leipzig, 1880.

Chace, A. B.; Bull, L.; Manning, H. P.; and Archibald, R. C.
The Rhind Mathematical Papyrus, Mathematical Association of America, Vol. 1 (1927), Vol. 2 (1929), Oberlin, Ohio.

Chronique D'Egypte
Bulletin Périodique de la Fondation Égyptologique Reine Élisabeth, publié avec le Concours du . . . la Fondation Universitaire de Belgique, Vol. 41, No. 81, Brussels, 1966.

Cottrell, L.
Life Under the Pharaohs, Pan Books, London, 1964.

Cottrell, L.
The Lost Pharaohs, Evans, London, 1949.

Cottrell, L.
The Mountains of Pharaoh, Pan Books, London, 1964.

Cottrell, L.
The Penguin Book of Lost Worlds, (2 volumes) Penguin, London, 1966.

Crum, W. E.
Catalogue of the Coptic Manuscripts in the British Museum, London, 1905.

Crum, W. E.
Coptic Ostraca from the Collections of the Egypt Exploration Fund, the Cairo Museum and Others, London, 1902.

Davidson, D., and Aldersmith, H.
The Great Pyramid, its Divine Message, London, 1924. Reprinted 1937.

Desroches-Noblecourt, Christiane
Egyptian Wall-Paintings From Tombs and Temples, Fontana Unesco Art Books, Collins, Unesco, Milan, 1962.

Driver, G. R.
Semitic Writing from Pictograph to Alphabet (British Academy Schweich Lectures, 1944), Oxford University Press, London, 1948.

Edgar, J. and M.
The Great Pyramid Passages and Chambers (2 volumes), Glasgow, 1910. Reprinted 1923.

Edwards, I. E. S.
The Pyramids of Egypt, Pelican Books, London, 1947.

Egyptian Mathematical Leather Roll (EMLR)
See S. R. K. Glanville, "The mathematical leather roll in the British Museum," *Journal of Egyptian Archaeology*, Vol. 13 (1927), pp. 232–238.

Eisenlohr, A.
Ein Mathematisches Handbuch der Alten Ägypter, (*Papyrus Rhind des British Museum*), *übersetzt und erklärt*, Leipzig, 1877.

Encyclopaedia Britannica
"Egypt-Mathematics," by F. L. Griffith, 11th edition, Vol. 9 (1910), pp. 46–47.

Engelbach, R.
"Mechanical and Technical Processes," in *The Legacy of Egypt*, S. R. K. Glanville, editor, Oxford University Press, London, 1942. Reprinted 1963.

Erman, Adolf
The Ancient Egyptians, A Sourcebook of their Writings, Introduction by W. K. Simpson, Harper Torchbooks, Harper, New York, 1966. First published 1923.

Eves, Howard
An Introduction to the History of Mathematics, Holt, Rinehart and Winston, New York, 1964. First published 1953.

Finger, Charles J.
Pepys' Diary, Haldeman-Julius Company, Girard, Kansas, 1922.

Gardiner, Sir Alan
Egyptian Grammar, being an Introduction to the Study of Hieroglyphics, Oxford University Press, London, 1957. First published 1927.

Gardiner, Sir Alan
Egypt of the Pharaohs, Oxford University Press, London, 1964. First published 1961.

Gillain, O.
La Science Égyptienne: l'Arithmétique au Moyen Empire, Paris and Brussels, 1927.

Gillings, R. J.
"An Interesting Ostracon," *The Australian Mathematics Teacher*, Vol. 19, No. 1 (Sydney, 1963), pp. 15–19.

Gillings, R. J.
"Mathematical Fragment from the Kahun Papyrus," *The Australian Journal of Science*, Vol. 29, No. 5 (November, 1966), pp. 126–130.

Gillings, R. J.
"Mathematical Fragment from the Kahun Papyrus IV, 3, *The Australian Mathematics Teacher*, Vol. 23, No. 3 (Sydney, 1967), pp. 61–64.

Gillings, R. J.
"Perpetual World Calendar," *The Australian Mathematics Teacher*, Vol. 1, No. 1 (Sydney, 1945), p. 24.

Gillings, R. J.
"Problems 1–6 of the Rhind Mathematical Papyrus," *The Mathematics Teacher*, Vol. 55, No. 1 (Washington D.C., 1962), pp. 61–69.

Gillings, R. J.
"Sum of *n* Terms of an Arithmetical Progression in Ancient Egypt," *The Australian Journal of Science*, Vol. 31, No. 1 (Sydney, 1968), pp. 47–50.

Gillings, R. J.
"The Addition of Egyptian Unit Fractions," *Journal of Egyptian Archaeology*, Vol. 51, (London, 1965), pp. 95–106.

Gillings, R. J.
"The Area of the Curved Surface of a Hemisphere in Ancient Egypt," *The Australian Journal of Science*, Vol. 30, No. 4 (Sydney, 1967), pp. 113–116.

Gillings, R. J.
"The Egyptian Mathematical Leather Roll (B.M. 10250)," *The Australian Journal of Science*, Vol. 24, No. 8 (Sydney, 1962), pp. 339–344.

Gillings, R. J.
"The Egyptian $\frac{2}{3}$ Table for Fractions," *The Australian Journal of Science*, Vol. 22, No. 6 (Sydney, 1959), pp. 242–250.

Gillings, R. J.
"The Remarkable Mental Arithmetic of the Egyptian Scribes," *The Mathematics Teacher* (Washington, D.C., 1966). Part I: Vol. 59, No. 4, pp. 372–381. Part II: Vol. 59, No. 5, pp. 476–484.

Gillings, R. J.
"The Volume of a Cylindrical Granary in Ancient Egypt," *The Australian Mathematics Teacher*, Vol. 22, No. 1 (Sydney, 1966), pp. 1–4.

Gillings, R. J.
"The Volume of a Truncated Pyramid in Ancient Egyptian Papyri," *The Mathematics Teacher*, Vol. 57, No. 8 (Washington D.C., 1964), 552–555.

Gillings, R. J.
"'Think of Number' Problems 28, 29 of the Rhind Mathematical Papyrus (B.M. 10057–8)," *The Mathematics Teacher*, Vol. 54, No. 2 (Washington D.C., 1961), pp. 97–100.

Gillings, R. J. and Rigg, W. J. A.
"The Area of a Circle in Ancient Egypt," *The Australian Journal of Science*, Vol. 32, No. 5 (Sydney, 1969), pp. 197–200.

Glanville, S. R. K. (editor)
The Legacy of Egypt, Oxford University Press, London, 1963. First printed 1942.

Glanville, S. R. K.
"The Mathematical leather roll in the British Museum," *Journal of Egyptian Archaeology*, Vol. 13 (London, 1927), pp. 232–238.

Griffith, F. L.
"Egypt-Mathematics," *Encyclopaedia Britannica*, 11th edition, Vol. 9 (1910), pp. 46–47.

Griffith, F. L.
The Petrie Papyri: Hieratic Papyri from Kahun and Gurob (2 volumes), F. L. Griffith, editor, University College, London, 1897.

Hayes, W. C.
"Ostracon No. 153," *Ostraca and Name Stones from the Tomb of Sen-mut at Thebes*, Publication 15, The Metropolitan Museum of Art, New York, 1942.

Heath, Sir Thomas L.
The Thirteen Books of Euclid's Elements (2nd edition, 3 volumes), Cambridge University Press, London, 1926.

Herodotus
History, Translated by George and Sir Henry Rawlinson, Everymans, London, 1912.

Hogben, Lancelot
Mathematics for the Million, George Allen and Unwin, London, 1945. First published 1936.

Hutton, Charles
A Philosophical and Mathematical Dictionary (2 volumes), F. C. and J. Rivington, London, 1815.

Hutton, Charles
Recreations in Mathematics and Natural Philosophy. Translated from Montucla's Edition of Ozanam, revised by Edward Riddle, Thomas Tegg, London, 1840.

Iverson, Erik
The Myth of Egypt and its Hieroglyphs in European Tradition, G. E. C. Gadd, Copenhagen, 1961.

Jaffe, Bernard
Men of Science in America, Simon and Schuster, New York, 1944 (p. 93).

James, T. G. H.
Egyptian Sculptures, Fontana Unesco Art Books, Collins, Milan, 1966.

Jourdain, Philip E. B.
"The Nature of Mathematics," in *The World of Mathematics*, Vol. 1, James R. Newman, editor, Simon and Schuster, New York, 1956.

Kahun Papyrus (KP)
The British Museum, London. Listed under: Kahun Papyri IV.2, IV.3, XLV.1, LV.3, and LV.4.

Kleppisch, K.
Die Cheopspyramide, ein Denkmal mathematischer Erkenntnis, Munich and Berlin, 1921.

Kline, Morris
Mathematics, A Cultural Approach, Addison-Wesley, Reading, Mass., 1962.

Life Science Library
Mathematics, by David Bergamini and the Editors of Life. Time-Life International, the Netherlands, 1965.

Mansion, P.
"Sur une table du Papyrus Rhind," *Annales de la Société Scientifique de Bruxelles*, Vol. 12 (Louvain, 1888), p. 46.

Michigan Papyrus (Mich.P)
University of Michigan Collection, Vol. III, Papyri Nos. 146, 621. See also L. C. Karpinski in *Isis*, Vol. 5 (Smithsonian Institute, Washington, D.C., 1923), pp. 20–25.

Milliken, E. K.
Cradles of Western Civilization, Harrap, London, 1955.

Moscow Mathematical Papyrus (MMP)
Moscow Museum of Fine Arts. No. 4576 of the Inventory.

Nairtz, O.
"Die Cheopspyramide, ein Viertausendjähriges Räthsel," *Prometheus*, Vol. 17 (1906).

Neikes, H.
Der goldene Schnitt und die "Geheimnisse der Cheopspyramide," Cologne, 1907.

Neugebauer, O.
The Exact Sciences in Antiquity, Harper Torchbooks, Harper, New York, 1962. First published 1951.

Neugebauer, O.
Vorgriechische Mathematik (Vorlesungen), Springer, Berlin, 1934.

Neugebauer, O.
"Zur ägyptischen Bruchrechnung," *Zeitschrift für Ägyptische Sprache*, Vol. 64 (Leipzig, 1929), p. 48.

Neugebauer, O. and Parker, R. A.
Egyptian Astronomical Texts I, Brown University Press and Lund Humphries, London, 1960.

Newman, James R.
"The Rhind Papyrus," in *The World of Mathematics*, Vol. 1, James R. Newman, editor, Simon and Schuster, New York, 1956.

Paterson, W. E.
School Algebra, 3rd edition, Clarendon Press, Oxford, 1916.

Peet, T. Eric
"A Problem in Egyptian Geometry," *Journal of Egyptian Archaeology*, Vol. 17 (London, 1931), pp. 100–106.

Peet, T. Eric
"Mathematics in Ancient Egypt," *Bulletin of the John Rylands Library*, Vol. 15, No. 2 (Manchester, 1931).

Peet, T. Eric
Review of Struve's Translation of the Moscow Mathematical Papyrus, *Journal of Egyptian Archaeology*, Vol. 17 (London, 1931).

Peet, T. Eric
The Rhind Mathematical Papyrus, British Museum 10057 and 10058, London, 1923.

Petrie, W. M. F.
Illahun, Kahun and Gurob, London, 1891.

Petrie, W. M. F.
The Pyramids and Temples of Gizeh, London, 1883.

Recorde, Robert
The Grounde of Artes. Teachyng the Worke and Practice of Arithmetike, 1542. See the *Mathematical Gazette*, Vol. 14, No. 195 (G. Bell and Sons, London, July, 1928), p. 196.

Reisner Papyri (RP)
The Museum of Fine Arts, Boston, Mass. Museum No. 38.2062.

Sanford, Vera
A Short History of Mathematics, Harrap, London, 1930.

Sarton, George
The Study of the History of Mathematics, Dover, New York, 1957. First published 1936.

Schack-Schackenburg, H.
"Das Kleinere Fragment des Berliner Papyrus 6619," *Zeitschrift für Ägyptische Sprache*, Vol. 40 (1902), p. 65.

Schack-Schackenburg, H.
"Der Berliner Papyrus 6619," *Zeitschrift für Ägyptische Sprache*, Vol. 38 (1900), p. 135.

Scott, A. and Hall, H. R.
"Laboratory Notes: Egyptian Leather Roll of the Seventeenth Century B.C.," *British Museum Quarterly*, Vol. 2 (1927), p. 56.

Sethe, Kurt H.
"Von Zahlen und Zahlworten bei den Alten Ägyptern," *Schriften der wissenschaftlichen Gesellschaft in Strassburg*, Part 25 (Trübner, Strassburg, 1916).

Sewell, J. W. S.
"The Calendars and Chronology," in *The Legacy of Egypt*, S. R. K. Glanville, editor, Oxford University Press, London, 1963.

Simpson, W. K.
The Papyrus Reisner I (1963), II (1965), III (1969). The Museum of Fine Arts, Boston, Mass.

Sloley, R. W.
"Science," in *The Legacy of Egypt*, S. R. K. Glanville, editor, Oxford University Press, London, 1963. First published 1942.

Smyth, Piazzi
Our Inheritance in the Great Pyramid, Daldy, Isbister, London, 1877. First ed. 1864; 2nd ed. 1874; 3rd ed. 1877; 4th ed. 1880; 5th ed., abridged, 1890.

Struik, Dirk J.
A Concise History of Mathematics, Dover, New York, 1948.

Struve, W. W.
"Mathematischer Papyrus des Staatlichen Museums der Schönen Künste in Moskau," *Quellen und Studien zur Geschichte der Mathematik*, Part A, Vol. 1 (Berlin, 1930).

Thompson, D'Arcy W.
On Growth and Form, Vol. 1 (1951), Vol. 2 (1952), Cambridge University Press, London.

Thompson, H.
"A Byzantine table of fractions," *Ancient Egypt* (London, 1914), p. 52.

Tsinserling, D. P.
"Geometry in Ancient Egypt," *Bulletin de l'Académie des Sciences de l'Union des Republiques Soviétiques Socialistes*, Vol. 19 (Leningrad, 1925), p. 541.

Turaev, B. A.
"The volume of the truncated pyramid in Egyptian mathematics," *Ancient Egypt* (London, 1917), p. 100.

Turnbull, H. W.
The Great Mathematicians, Methuen, London, 1951. First published 1929.

Van der Waerden, B. L.
"Die Entstehungsgeschichte der ägyptischen Bruchrechnung," *Quellen und Studien zur Geschichte der Mathematik*, Part B, Studien IV (Berlin, 1937–1938), p. 359.

Van der Waerden, B. L.
Science Awakening, English translation by Arnold Dresden, Noordhoff, Groningen, 1954.

Vogel, Kurt
"Eweitert die Lederrolle unsere Kenntniss ägyptischer Mathematik?" in *Archiv für Geschichte der Mathematik, der Naturwissenschaften und der Technik*, Julius Schuster, Berlin, 1929, p. 386.

Vogel, Kurt
"The Truncated Pyramid in Egyptian Mathematics," *Journal of Egyptian Archaeology*, Vol. 16 (London, 1930), p. 242.

Vogel, Kurt
Vorgriechische Mathematik, Vol. 1, *Vorgeschichte und Ägypten*, Schroedel, Hannover, 1958.

Wheeler, N. F.
"Pyramids and their Purpose," *Antiquity*, 9 (Gloucester, England, 1935).

Wilson, John A.
Signs and Wonders upon Pharaoh: A History of American Egyptology, The University of Chicago Press, Chicago, 1964.

Yeldham, F. A.
The Teaching of Arithmetic Through 400 Years, Harrap, London, 1936.

INDEX

Abacus, 11
Abd-el-Rasoul, 246
Abundant numbers, 178n
Abu Simbel, 90
Addition
  examples of in hieratic, 10, 12, 14, 15
  hieroglyph for, 6
  how the scribe did it, 13, 15
  possible tables for. See Tables
  scribes' ability in, 11
Aha (quantity) problems, 158, 181
A'h-mosè (Ahmes), the scribe, 45, 76,
  131–132, 139, 141–142, 145, 148
  falters with some fractions, 160
Akhenaton, Pharaoh, 90
Akhmîm Papyrus (AP or AMP), 28, 91
Akkadian cuneiform script, 7
Akkadians, 91
Aldersmith, H., 239
Alternando, 135–136
Amenhotep I, King, 220
Amosis, King, 82
Anklet, weight of, 212
Apries, Pharaoh, 207
Arabic, direction of writing, 5
Arabs, 2, 47
Archibald, R. C., 3n, 6n, 153, 217, 239
Archimedes, 200
Areas
  of circles, 139–140, 243
  of triangles and rectangles, 137, 243,
  246–247
  of hemispheres. See Hemisphere
Arithmetic progressions
  examples in RMP, 243–244
  properties of, 170
  terms in descending order of
  magnitude, 170, 174
  with common difference 5½, 171
Arura
  areas of fields in, 217
  definition of, 209

Assyriologists, 2

Babylon, 2, 240
Babylonian
  clay tablet (Plimpton 322), 1
  cuneiform script, 7
  direction of writing, 5, 6
Babylonians, 91, 240
Baskets
  conventional shapes for Egyptian,
  200
  half-an-egg, 195
  hieratic word nbt for, 194
  of grain, 90-hinu, 163
Becker, O., 48
Beer
  des-jugs of, 212
  division of. See Fractions
  strength of. See Pesu
  two types distinguished, 124, 125
Bergamini, David, 6n
Berlin Papyrus 6619 (BP), 91, 161n,
  216
  hieroglyphic sign for square root,
  216
Besha, 128
Bible, 237
Birds, and direction in which hieroglyphs face, 6
Bobynin, V. V., 48
Borchardt, L., 124–125
Boyer, Carl B., 139
Bruins, E. M., 48, 70
Bull, L., 3n, 6n
Burroughs, Edgar Rice, 237
Bushel, 147, 210
Byzantine
  tables of fractions, 91
  times, 16, 47

Calendar, Egyptian
  only intelligent one in human

Calendar, Egyptian (*continued*)
  history, 235, 240
  solar year of 365¼ days, 235
Canon for decompositions in the
  RMP, 49. *See also* Precepts; Recto
  Table of the RMP
Carter, Howard, 237
Chace, Arnold B., 3, 48, 109, 128n,
  133, 141–142, 153, 168, 181, 183,
  210, 212–213, 249
Champollion, Jean François, 1
Check marks, 18, 19, 20, 22n
Cheops (Khufu), pyramid of, 91, 231,
  237, 238
Chronology, 260–265
Circles, area of. *See* Areas
Collignon, E., 48
Coming forth (growing) period,
  235
Computers
  KDF-9, 50
  Use of series in, 19
Copernicus, 235
Coptic
  inscriptions, 86
  times, 16
Cottrell, Leonard, 239
Crum, W. E.
  catalogue of, 29
  table of fractions, 60n
Cubit
  cubic, 163–165
  Greek, 207
  Roman, 207
  royal, 207–208, 220
  short, 207–208
  strip, 137, 209
  superficial, analogy with superficial
    foot, 227–229
Cuneiform writings, 1, 5
Cylinder, area of curved surface of,
  194

Dattel, 128
Davidson, D., 239
Davidson, John, 239

Day, origin of division into 24 hours,
  236
Deben, definition of, 212
Decan, 236
Decimal system, 4, 12
Degree of civilisation, 3
Deir el-Bahri, temple of, 86, 246
Demotic writings, 2, 5
Descartes, René, 1
Des-jug
  of beer, 128, 212
  quantity contained in, 213
  See also Beer
Dictionnaire Égyptienne, 1
Digit. *See* Finger
Dinostratus, 198, 200
Diophantus of Alexandria, 181
Direction of ancient Egyptian writ-
  ing, 6
Dividendo, 135–186
Division
  by the clerk of the Reisner Paypri,
    219, 226
  of loaves among workmen, 120
  of loaves in various proportions,
    171–173
  of loaves, justice appears to be done,
    105
  of particular numbers: 184 by 8, 19–
    20; 1060 by 12 and 159, 252
  of nine loaves among ten men, 47,
    105n
  a powerful technique with fractions,
    204
  by ten, 27, 219, 223
  used to solve equations, 159
"Do it thus" occurring at the end of a
  problem, 183
Dot indicating hieratic fractions,
  21
Doubling of 1/10 to obtain 1/5, 84, 85
Drachmas and fractions, 47
Dresden, Arnold, 191n, 232n

Edgar, J. and M., 239
Edwards, I. E. S., 207n, 239

Egyptian Mathematical Leather Roll
(EMLR), 12, 22, 23n, 39, 89, 106,
109
    author of, 89
    dimensions of, 89
    early opinions of, 90
    entries with like terms, 111
    error in line 17 of, 99
    grouping of equalities in, 95
    photograph of, 94
    reconstruction of, 92
    three- and four-term equalities, 114
    translation of, 93
    unrolling of, 89
Egyptian standard tables, 95, 110
Egyptian symbols for numbers, 5
Egyptologists, 90–91, 246
Egypt, map of, 266
Eisenlohr, A., 48, 210
Emmer, 128
Ellen (cubits), 188
Englebach, R., 191
Enigmatic writing in the RMP, 245
Enlistees (workmen), 218–219, 224
Epagomenal ("year-end") days, 235
Equals, the hieratic sign nearest to it.
    See "This is"; Hieratic characters
Equations, solution of
    in MMP, 247
    in RMP, 243
Errors
    in addition and subtraction, 11
    infrequent in the Recto of the
    RMP, 47
    in multiplications involving finger-
    breadths, 229
    found in the Reisner Papyri, 222
Even number fractions, 71–73. See
    also Odd and even numbers in the
    RMP Recto
Eves, Howard, 240
Exact Sciences in Antiquity, 11n, 88n,
    235n, 240n, 241
"Examples of proof," the hieratic
    form of this phrase, 9
Exchange of loaves

of different pesus, 129, 130, 132–133;
    using modern algebraic techniques
    for, 134, 136
Eyth, M., 239
Ezekiel, 207

False assumption (false position),
    method of, 154, 157, 162, 243
Favarro, A., 48
"Find," the hieratic sign for, 9
Finger, Charles J., 17n
Finger (fingerbreadth)
    definition of, 208
    inclusion of, in measurements at the
    dockyard workshops, 228–229
    used in calculation of sekeds, 187
Flinders, Matthew, 91n
Foundations of Egyptian mathe-
    matics, 3
Fractions
    Egyptians unable to write the com-
    mon fraction p/q, 234 and
    fractional divisions of bread and
    beer, 126–127
    hieroglyph for, 20
    Horus-eye, 173n, 210–211
    meaning of, to scribes, 110
    multiplication of, 224
    no two alike together, 95n, 105–107.
    See also Egyptian Mathematical
    Leather Roll, grouping of equalities
    possible hieroglyph for ¾, 20n
    problems involving, genesis of, 105
    scribe flounders in a maze of, 160
    series of, in descending order of
    magnitude, 102n
    sign for, in hieratic, 21
    with very large denominators, 159

Gallons, 147, 210
Gardiner, Alan
    Egyptian dictionary of, 165
    Egyptian Grammar, 207n
    on the Reisner papyri, 218n
Generators
    definition of, 43, 104

Generators (*continued*)
examples of, 108, 111–113
for pairs of fractions, 33, 43–44
table of, 116
Geometric progression
unique properties of, 167
with 7 the common ratio, 168, 244
Gillain, O., 48
Gillings, R. J., 176n, 177n, 181n
Glanville, S. R. K., 16n, 89–91, 98,
103, 145n, 207n
Golden section (sectio aurea), 238
Golenischev, V. S., 91n, 246. *See also*
Moscow Mathematical Papyrus
Gradient. *See* Seked
Granaries, 2, 146–148
rectangular, 163–164
volumes of, 243
Great Sphinx, 237
Greek
papyrus, 217
tables of fractions, 91
times, 16
Greeks, 2, 47, 232, 234, 240
Griffith, F. Ll., 48, 73–74, 148n, 152,
156n, 162, 176, 216
Grotefend, Georg F., 1
Grounde of Artes, 17
G rule
beginnings of, 95–96
examples of, 186, 215
an exercise in, 42
extension of, 41
further extension of, 44
not explicitly stated in papyri, 39
probable use of, 41
Gunn, B. G., 189, 191, 210

Hall, H. R., 89, 90, 103
Hamblin, C. L., 50n
Handbreadth. *See* Palm
Harmonic mean, the concept of, 131,
244
Hatshepsut, Queen, 86
Hayes, W. C., 87n
Hayt, 162–165, 208

Head lay priest of the temple,
126
Head reader of the temple, 126
Heap, an unknown quantity, 157
Hekat (heqat)
definition of, 210
measure of grain, 128n, 151, 163,
165, 202, 253
Hemisphere, area of surface of, 188,
194, 195, 196, 198, 200, 246
Herodotus, 238-240
Hesy-Ra, panel from tomb of, 83
Hieratic
characters, 4–10
integers from various papyri, 255–
256
signs for khets and hekats, 141
sign "this is," meaning equals, 104,
123
variations in writing fractions, 259
variations in writing integers, 257–
258
Hieroglyphs
decipherment of, 1
direction in which pictographs face,
6
for numbers, 4, 5
for two-thirds, 21
invention of, 4–6
written from right to left, 5
Hilfszahlen, 81n
Hinu, 163–165
definition of, 210
Hogben, Lancelot, 48
Hophra. *See* Apries, Pharaoh
Horus, Egyptian God, 210, 235
Horus-eye fractions, 173n, 202, 204,
206
for grain, 210, 244, 247, 249, 253
hieroglyphic and hieratic signs for,
211
in terms of hinu, 250
use of, in the RMP, 243
Hultsch, F., 48
Hutton, Charles, 181–182
Hyksos period, 18, 45

Ibscher, H., 218
Illahun, temple of, 124, 231
Illegible signs, how indicated, 7
Index laws of algebra, 1
Introduction to the RMP by
  A'h-mosè, 183
Inundation (sowing) period, 235
Inverse proportion, 244
Irrigation canals, 2
Isis, Egyptian Goddess, 210, 235
Isosceles triangle, area of, 139

James, T. G. H., 194
Jourdain, P. E. B., 16, 232
Justice appearing to be done in the
  division of loaves, 105

Kahun Papyrus (KP), 39, 91
  portion of the RMP Recto included
    in KP IV 2, 104, 176
  Problem KP IV 3 (contents of a
    granary), 148, 151
  Problem KP LV 4 (rectangular
    granary), 162–165, 216
  Problem KP LV 3 (mathematical
    fragment), 156
  six mathematical items, not all
    penetrated yet, 176
Karnak, temple of, 2, 90, 237
KDF-9 computer, 50n
Khar
  contents of granaries in, 147–148,
    152
  definition of, 146, 151, 210
  values in, 163, 165
Khet
  measure of length, 137–139
  definition of, 209
  hieratic signs for, 141
Kleppisch, K., 239
Kline, Morris, 175, 240
Köbel, Jacob, 181–182

Largest number in the Recto of the
  RMP, 49n, 67
Least common denominators, 78, 81

table of common measures, 251
Life Science Library: Mathematics, 6
Lindgren, Harry, 146n, 191
Loaves of Egyptian bread, 212, 244
Logarithms, 1
Loria, G., 48
Lotus, 5
Luckey, P., 191
Luxor, temple of, 2, 89

Manning, H. P., 3n, 6n
Mansion, P., 48, 70
Mathematical papyri, 6
Measuring reed, or rod, 219–220
Mental arithmetic of scribes
  evidence of, 19, 248, 251–252
  further illustrations of, 159–160, 186,
    199, 224, 226
  impossibility of, in Problem 70 of
    the RMP, 107
Meret (side or height of a triangle),
  138n, 139
Meryre, the official, 82
Michigan Papyrus (Mich. P.), 91
Middle Kingdom, 124, 175
Middle Kingdom hieratic, 7, 18
Milliken, E. K., 86n
Montucla, 181
Moscow Mathematical Papyrus
    (MMP), 39, 91, 104, 246–247
  Problem 6, reference to square root,
    216
  Problem 10, 194
  Problem 19, 157–158
  Problem 21, 131–132
Mother Goose rhyme and Problem 79
  of the RMP, 15, 169
"Mouth," hieroglyphic word tp-r for,
  194
Multiplication
  by continued doubling, 166
  examples of, 17–19, 204, 219, 221–
    223
  form of, devised for the clerk of the
    dockyard workshops, 227, 230
  of fractions, 22–23, 225

Multiplication (*continued*)
  hieratic signs for, 8, 10
  of one-fifth by ten, 22
  since the times of ancient Egypt, 16
  in the Sixteenth and Seventeenth
    Centuries, 17
  by ten, 13–14, 21, 23, 27n
Multi-term fractional equalities, 115

Nag'ed Deir
  dockyard workshops of, 218, 229
  Reisner papyri found at, 218
Nairtz, O., 239
Napier, John, 1
Nebuchadnezzar, King, 207
Neikes, H., 239
Nephthys, Egyptian God, 235
Neugebauer, O., 1, 11n, 88, 90, 103,
    109, 194, 235, 240
Newman, J. R., 16, 232–233
Nile River
  annual flooding of, 235
  clay from, 189
  papyrus reeds from, 2, 4
Noetling, F., 239
Number systems, readable, 7
Numbers, very large, 219

Octagon
  inscribed in a square, 143
  regular, 142
Odd and even numbers in the Recto
  of the RMP, 49
One-third
  of any number, 25–26
  of mixed numbers, 35
  of one and three, 34, 155n
  of 30, finding two-thirds first, 184
Osiris, Egyptian God, 210, 235
Ostracon No. 153 from the tomb of
  Sen-mut, 86–88
Ozanam, 181

Palette as scribe's insignia, 81, 82
Palm, 185, 207
Paper, 4

Papyri
  Greek, 209
  hieratic, 2
Papyritic memo pad, 19, 22, 107, 186,
    203, 219, 248, 251
Papyrus reed, 4
Paterson, W. E., 175n
Peet, T. Eric, 11n, 24, 48, 139, 145–
    146, 183, 189, 191, 194–198, 200,
    210, 213, 216, 232
Pepys, Samuel, 17
"Perpetual World Calendar," 235
Persians, 2
Pesu
  definition of, 128, 212
  of bread and beer, 128, 240
  exchange of loaves using modern
    algebra, 134–136
  loaves of different pesus exchanged,
    129–133
  meaning of, 128n
  various modern terms for, 213
Petrie Papyri, 148n, 152n, 156n,
    162n
Petrie, William Matthew Flinders,
    91n
Phrases, mathematical, in hieratic, 7
Pi (π), the Egyptian equivalent $256/81$
  for, 142, 197, 199, 238, 240
Plimpton 322 (Babylonian clay
    tablet), 1
Potsherds, 86
Precepts for the RMP Recto fractions,
    49
Prime equalities, 115
Problems
  in completion, 81
  RMP 28 on addition and subtrac-
    tion, 6n
Profiles in hieroglyphs, 6
Proof, on the nature of, 145–146,
    232
Ptolemaic demotic writing, 7
Ptolemies, 209
Puttock, M. J., 220
Pyramidiot, The Great, 239

Pyramids
  dimensions of the Giza group of, 185n
  dissection of, 189, 190
  juel type of, 190
  orientation of, 186
  Problems RMP 56–60 and MMP 14 on, 185–189
  right, 185
  scribes' drawings of, 186, 188
  standard formula for a frustum, 189, 192, 193, 246
  truncated form of, 185, 187, 188, 193, 247
  volume of, 185, 189
Pythagorean theorem
  in ancient Egypt, 242
  known to the Babylonians, 1
  suggestions of, in equations, 161
  unknown to Egyptians, 238

Quadruple hekat, 151, 210
Quedet (weight of finger ring), 212

Rameseum, 89
Ramses, Pharaoh, 90
Rarity of scribal errors, 11. *See also* Errors
Reciprocal of 1½ is two-thirds, 21, 27, 157, 158
Recorde, Robert, 17
Rectangle, area of, 137
Recto Table of the RMP, 22, 34, 109, 243
  A'h-mosè falters on 2 ÷ 95, 68
  canon or precepts for the values of, 49
  derivation of some of the values of, 98
  lack of scientific attitude claimed, 232
  method for 2 ÷ 35 not standard, 79
  most extensive of all extant tables, 45
  simplest and best values, 47
  a whole new one possible, 69

Red auxiliaries, 78, 81, 85, 87, 97, 99, 102–103, 160–161, 251–252
Reference numbers for the red auxiliaries, 97, 101, 103, 251–252
"Regular share," meaning of the hieratic phrase, 173
Reisner Papyri, 91, 218
  the seventeen divisions of, 219
"Remainder," hieratic sign for, 8
Remen
  definition of, 208
  and double-remen, 208
  use in doubling areas, 208
Reptiles, directions in which hieroglyphs face, 6
Rhind, A. H., 89
Rhind Mathematical Papyrus (RMP), 3, 3n, 21, 24, 39, 73, 90, 154
  details of the contents of, 243–245
  dimensions of, 45
  mechanical arithmetic of, 91
  scribe's introduction to, 45
Rigor, conditions for, of a nonsymbolic proof, 233
Romans, 47
Rosetta Stone, 7
Ro
  definition of, 210
  hieroglyphs and hieratic signs for, 211
  unit of measure, 47, 129, 202, 205
Royal cubit, 139

Sachs, A., 1
Sacrificial bread, 247
Saint Andrew's cross in multiplication, 17
Sanford, Vera, 181
Scales of notation
  scale of 4, 230
  scale of 7, 227–228
Schack-Schackenburg, H., 48, 148, 153, 161, 176, 216
Scheffel, 210
Schoenia, 209
Science, as the Egyptians knew it, 234

Scott, Alexander, 89, 90, 103
Scribes
  accuracy in multiplication, 222–223
  directions in which hieroglyphs
    face, 6
  handwriting of, 4, 196
  hieroglyphs for, 81
  never verbose, 7, 168
  preoccupations with fractions, 104
  of the temple, 126
Seked
  calculation of, 185
  and cotangent of angle, 187, 212
  definition of, related to gradient,
    212
  of Giza pyramids, 187
  of juel pyramids, 190–191
  and slope of the sides of pyramids,
    185, 187, 244
Semicircle
  area of, 196, 198
  circumference of, 197
Semicylinder, area of curved surface
  of, 196–198, 200
Sen-mut, tomb of, 86–87
Series
  arising from ordered 2-term
    equalities, 119
  a special property of, for addition,
    19
  used in multiplication, 166
Sesostris I, Pharaoh, 218
Setat (square khet), 38–39, 140–141,
  143
  definition of, 209
Seth, Egyptian God, 210, 235
Sethe, K. H., 210
Seven, the number, 99, 168–169. See
  also Scales of notation.
Sewell, J. W. S., 236n
Sexagesimal system, 4
Shapes and Sections, 149n
Shaty, definition of, 212
Simon, M., 48
Simpson, W. K., 218, 223
Simultaneous equations, one of the

second degree, 162
Sirius (the Dog Star), heliacal rising
  of, 235–236
Six-sevenths, as written in Egyptian
  fractions, 21
Sloley, R. W., 16, 145, 207n, 210
Smither, Paul C., 218n
Smyth, Piazzi, 237
Snakes, directions in which hieroglyphs face, 6
Spelt-dates, 128
Square root
  approximations to, 215
  hieratic word and signs for, 162–163,
    176
  hieroglyphic signs for, 216
  of particular numbers: $1\frac{1}{2}$ $\frac{1}{16}$, 161–
    162; 164 in approximation, 217;
    $6\frac{1}{4}$, 162
  possible tables for, 214–215
  simple cases of, 138, 143, 161–162,
    164
Squaring numbers, 214
  the fraction $\frac{1}{2}$ $\frac{1}{4}$, 161–162;
  the mixed number $8\frac{2}{3}$ $\frac{1}{6}$ $\frac{1}{18}$, 147
  the numbers 2 and 4, 188
Squaring the circle: A'h-mosè perhaps
  the first of the circle squarers, 145
Stammbruchen, 21
Stammbruchsummen, 90
Standard equality in fractions (the
  equality $\frac{1}{7}$ $\frac{1}{14}$ $\frac{1}{28}$ = $\frac{1}{4}$ of the
  EMLR), 100–102, 206, 209, 253
Standard fraction tables, 91, 110
Stewart, Edgar, 239
Struik, D. J., 16, 139, 232
Struve, W. W. 131n, 157n, 185, 194–
  195, 198, 210, 213, 216, 246–247
Subtraction
  Egyptian method for, 11
  examples of, in RMP, 243
  the hieroglyphic sign for, 6, 10
Sumerian cuneiform script, 7
Summer (harvest) period, 235
Sum to n terms of an arithmetic
  progression, 175

Superficial cubit, and superficial foot, 227–229
Sylvester, J. J., 48

Tables
of addition and subtraction in hieratic, 11–14
elementary forms of, for fractions, 111
of fractions, 11, 15, 22, 32
of fractions of a cubit, 209, 225
from geometric progressions, 168
in the RMP and elsewhere, 106
of squares, 162, 214–217
of two-thirds of numbers, 153
for use of functionaries of the granary, 249
Tadpole, 5
Tannery, P., 48
Tarzan, 237
Taxes, 3
Taylor, John, 239
Tell el 'Amarna, clay tablets from, 91
Temples, 2, 231, 237, 246
Teper (base or mouth of a triangle), 138n, 139, 194
Thales, 2
Thebes, 87, 89
"The doing as it occurs," the hieratic form of this phrase, 9
Theory of numbers, faint beginnings of, 96, 114, 119
"Therefore," the hieratic form of this word, 9
"Think of a number" problems, 158, 181–183, 243
This (Egyptian town), 218
"This is," hieratic sign for
occurring in each of RMP Problems 1–6 and 87 times altogether, 123
occurs 50 times in EMLR, 92, 104
Thomas, W. R., 191
Thompson, D'Arcy W., 238
Thoth, Egyptian God, 210
Three-quarters as written in Egyptian fractions, 21

Three-term unit fractional equalities, 108
"Total," hieratic sign for, 9, 15
Transliteration, hieratic to hieroglyphics, 7
Triangle, area of, 138
Tsinserling, D. P., 246
Turajeff, B. A., 246
Turnbull, H. W., 238
Tutankhamen, Pharaoh, 237
Tuthmosis IV, Pharaoh, 82
Twice-times table, 3, 234
Two divided by
forty-five, 51, 52
multiples of 3, alternative methods, 77
thirteen, 71, 79–80
thirty-five, scribal method disclosed, 77–78
Two-thirds, 20n, 21
Two-thirds of any number
examples and technique, 24
in later times, 28–29
rule for even fractions, 38
rule for odd fractions, 29
scribes' ability to find, 3, 234, 244
Two-thirds of the fraction $\frac{1}{169}$, 160
Two-thirds tables
for any number, routine, 153
for integers and fractions, 23
possible method of constructing, 24
some justification for, 95, 251

Unitary method, 130
Unit fractions
Egyptian, 21–23
used by Greeks two millenia later, 47
Usual reader of the temple, 126

Van der Waerden, B. L., 48, 90n, 103, 191, 194, 213, 232
Verso of the RMP, 22n
Vertical writing, 6
Vogel, Kurt, 48, 81n, 90, 97n, 103, 128n, 142, 191, 213

Volume
  of a cylindrical granary, methods for,
    146
  of a rectangular prism, 189
  rule for finding directly in khar, 148,
    150
  standard Egyptian rule for finding,
    150

Wedyet flour, 128, 130
"Working out," the hieratic form of
    this phrase, 9
Writing
  how it started, 4
  instruments used for, 82–83
  vertically downwards, 6

"You have correctly found it"
  the hieratic signs for this phrase, 10
  as used in the MMP, 157

Zero, indicated by a blank space, 228